Contents

Practice Tests for Communications Licensing and Certification Examinations
The Complete TAB Reference

Sam Wilson
Joseph A. Risse

TAB Books
Division of McGraw-Hill, Inc.

New York San Francisco Washington, D.C. Auckland Bogotá
Caracas Lisbon London Madrid Mexico City Milan
Montreal New Delhi San Juan Singapore
Sydney Tokyo Toronto

pbk   1 2 3 4 5 6 7 8 9 DOH/DOH 9 9 8 7 6 5 4

Library of Congress Cataloging-in-Publication Data
Wilson, J. A. Sam.
    Practice tests for communications licensing and certification
examinations : the complete TAB reference / by Sam Wilson, Joseph A.
Risse.
        p.   cm.
    Includes index.
    ISBN 0-07-070824-X (pbk.)
    1. Radio—Examinations, questions, etc.   2. Radio operators-
-Certification.   3. Electronics—Examinations, questions, etc.
4. Electronic technicians—Certification.   I. Risse, Joseph A.
II. Title.
TK6554.5.W49 1994
621.382'076—dc20   94-30237
                                                                CIP

Acquisitions editor: Roland S. Phelps
Editorial team: Joanne Slike, Executive Editor
                Andrew Yoder, Book Editor
Production team: Katherine G. Brown, Director
                Ollie Harmon, Coding
                Susan E. Hansford, Coding
                Rose McFarland, Layout
                Linda L. King, Proofreading
                Joann Woy, Indexer   EL1
Designer: Jaclyn J. Boone   070824X

Acknowledgments

The authors thank Roland Phelps, the TAB acquisitions editor in charge of this book, for his help in getting the book started and finished. The following people were also very helpful in getting the book into production: Joanne Slike, Andrew Yoder, and Kriss Lively-Helman.

In addition, we appreciate the help of Joseph Glynn and William Myers, both of WVIA TV, Pittston, PA. They provided valuable information on the subject of Television Transmission and power measurement.

Thanks also to Ken Muncey of Cocoa Beach, Florida for reviewing the rough draft and for his helpful suggestions.

A special thanks goes to Norma Wilson of Melbourne, FL and Lisa Naipaver of Cleveland, OH. They typed the rough drafts and the manuscript. Working together, they converted our ideas into a readable manuscript.

Electronic Servicing and Technology (ES&T) magazine gave us broad permission to use material from their publication. We appreciate it very much.

Several organizations gave much extra help with study material. We give special thanks to ETA, ISCET, and SBE for sending us a lot of useful material.

Introduction

Practical applications of electronics started with communications. Throughout the history of electronics, the field of communications has been a technological leader. Studying communications has always been an excellent way to learn about practical electronics.

A number of licenses and certificates are available to people with training in basic electronics and communications. Each is obtained by taking a test in some particular specialty. Following is a list of the certification and license exams that are available (the abbreviations are explained below).

- GROL
- GMDSS
- Radio telegraph certificate
- Radar endorsement
- Amateur radio technician class license
- Marine radio operator permit
- ETA associate-level certification
- ISCET associate-level certification
- ETA journeyman communications certification
- ISCET journeyman communications certification
- NICET certification
- SBE certification
- NARTE

GROL   General Radio Operator License
GMDSS   Global Marine Distress and Safety Service
ETA   Electronics Technicians Association
ISCET   International Society of Electronics Technicians
NICET   National Institute for Certification in Engineering Technologies
SBE   Society of Broadcast Engineers
NARTE   National Association of Radio and Telecommunications Engineers

There is a companion book for this study guide. It is titled *The TAB Sourcebook for Communications Licensing and Certification Examinations*. The two books give you an overall review of the material needed for passing required tests for licenses and certifications.

This is not a textbook. It is a book designed to give technicians who have had training and/or experience a chance to evaluate their ability to pass license and certification tests. It is designed to reveal subjects where further study would be useful before taking a test.

It is also a useful book for technicians who have recently completed their training in electronics. Because of the wide range of electronics subjects covered, technicians can find places where they would benefit from further study.

When you answer a question correctly, you reinforce your knowledge. When you cannot answer a question, you have a chance to increase your knowledge.

By showing a willingness to put knowledge to the test, technicians demonstrate that they can accept challenges in their work. Also, it demonstrates pride in their knowledge and ability to accept challenge. As an added incentive, for some jobs, a license and/or certification is required to obtain employment.

In order to cover the wide range of subject matter necessary, the authors have made a few assumptions. For example, it is assumed that you are preparing to take one or more of the tests required for licensing and/or certification. For some of those tests, but not all of them, it is necessary to have an understanding of basic algebra and trigonometry. It is not possible to include preparation for those subjects in this book.

It is also impossible to teach Morse code in a book. The amateur technician license we prepare you for does not require code. Code is required for the radiotelegraph licenses. So, we are only preparing you for the theory part of the required exam for that license.

Calculations for solved problems in this book are usually carried out to three significant figures.

Very little coverage is given to the review of very basic electricity and electronics in this book. It is assumed you are familiar with Ohm's law, series and parallel dc and ac circuits, and the theory of operation for transistors and other basic components and circuits.

Answers are given for every question used in this book, so it can be used for either classroom work or self study.

To keep in step with most of the license and certification tests, we assume electron flow to be the direction for current. If you have been trained in conventional (positive-to-negative) current flow, you should practice reversing the direction of current flow when studying for the required tests

When you add your existing knowledge to the material presented in these books, you should not experience any unpleasant surprises when you sit for the actual tests.

1
CHAPTER

Special categories in the GROL pool of questions

There are some subjects in the GROL pool of questions that do not fit well into the chapter headings of this book. You can expect to find questions on those subjects if you take a FCC. The questions in this book are worded exactly the way you will find them in the actual test. An exception is questions in the end-of-chapter quizzes.

Even though your training and experience might be outside the subjects in this chapter, you should be able to answer the questions on the FCC test after studying this material. If you have trouble with any of these subjects, it would be a good idea to do some outside reading to enhance the material given here.

Power equipment

1. The condition of a storage battery is determined with a:
    A. Hygrometer.
    B. Manometer.
    C. PET.
    D. Hydrometer.

The correct answer is D. A hydrometer measures the specific gravity of a fluid. In this case, the fluid is the battery electrolyte.

2. Here is your next question: The electrolyte in a lead storage battery is:
    A. Potassium hydrate.
    B. Pure spongy lead.
    C. Sulfuric acid.
    D. Iron oxide.

The correct answer is C. Actually, the electrolyte is a dilute solution of sulphuric acid and distilled water.

3. Here is your next question: The stage of charge of a lead-acid storage cell is determined by:

    A. Hydrometer reading.
    B. Its open circuit voltage.
    C. Its short circuit discharge current.
    D. Ohmmeter.

The correct answer is A.

4. Here is your next question: In storing a fully charged battery (lead-acid) for a long period of time (8 to 12 months) one should:

    A. Drain and flush out the electrolyte then refill with distilled water.
    B. Keep it in a warm place.
    C. Jar it at least once a week to prevent sulfation.
    D. Maintain a constant trickle charge.

The correct answer is A.

5. Here is your next question: How can alternator whine be minimized?

    A. By connecting the radio's power leads to the battery by the longest possible path.
    B. By connecting the radio's power leads to the battery by the shortest possible path.
    C. By installing a high-pass filter in series with the radio's dc power lead to the vehicle's electrical system.
    D. By installing filter capacitors in series with the dc power lead.

The correct answer is B.

6. Here is your next question: How can conducted and radiated noise caused by an alternator be suppressed?

    A. By installing a blocking capacitor in the field lead.
    B. By connecting the radio's power leads to the battery by the longest possible path and by installing a blocking capacitor in series with the positive lead.
    C. By installing a high-pass filter in series with the radio's power lead to the vehicle's electrical system and by installing a low-pass filter in parallel with the field lead.
    D. By connecting the radio's power leads directly to the battery and by installing coaxial capacitors in the alternator leads.

The correct answer is D. If you are familiar with the alternators used in cars, you know that the rectifying diodes are built into the alternator frame. That would make it difficult to perform the modification indicated in the answer. However, the question and answer are not relative to car alternators.

    Now, this is a very important point. Do not search for special cases that you might know of when answering FCC questions.

7. Here is your next question: A dynamotor is used to:

    A. Step up ac voltage.
    B. Step down ac voltage.
    C. Step down dc voltage.
    D. Step up dc voltage.

The correct answer is D. A dynamotor is a motor/generator set usually having the motor and generator mounted on the same shaft.

8. Here is your next question: Interference to radio receivers in automobiles can be reduced by:

    A. Connecting resistances in series with the spark plugs.
    B. Using heavy conductors between the starting battery and the starting motor.
    C. Connecting resistances in series with the starting battery.
    D. Grounding the negative side of the starting battery.

The answer given by the FCC is B. However, you know that resistor spark plugs have been used for many years to reduce ignition noise. To answer this question, assume that the resistor plugs are already installed.

9. Here is your next question: Normally in a mobile radio installation, the E output of a dynamotor is:

    A. Adjusted by armature rheostat.
    B. Adjusted by field rheostat.
    C. Not adjustable.
    D. Adjusted by battery rheostat.

The correct answer is C.

10. Here is your next question: If a shunt motor, running with a load, has its shunt field opened, how would this affect the speed of the motor?

    A. Slow down.
    B. Stop suddenly.
    C. Speed up then slow down.
    D. Unaffected.

The correct answer is C.

11. Here is your next question: The output voltage of a separately excited ac generator (running at a constant speed) could be controlled by:

    A. Changing the dielectric constant of the armature.
    B. Changing the material of the armature.
    C. The field current.
    D. Operating the generator in a no-load condition.

The correct answer is C.

12. Here is your next question: Static and interference from motors can be eliminated by:

    A. Grounding the battery with a 2" copper strip.
    B. Installing Faraday shields around connectors.
    C. Installing RF chokes across the power line to ground.
    D. Installing bypass capacitors from the power line to grounded parts

The correct answer is C. That is the answer supplied by the FCC.

13. Here is your next question: Auto interference to radio reception can be eliminated by:

    A. Installing resistive spark plugs.
    B. Installing capacitive spark plugs.
    C. Installing resistors in series with the spark plugs.
    D. Installing two copper-braid ground strips

The correct answer is A. Compare this question and answer with Question #8.

14. Here is your next question: A dc series motor speed is affected by:

    A. The load.
    B. The ripple frequency.
    C. The number of brushes.
    D. Stray RF fields

The correct answer is A.

15. Here is your next question: The output voltage of a separately excited ac generator with constant frequency is dependent upon the adjustment of:

    A. Iron or ferrite choke core.
    B. Field current.
    C. Brush position.
    D. Phase angle.

The correct answer is B.

16. Here is your next question: A dynamotor is approximately:

    A. 100% efficient.
    B. 85% efficient.
    C. 65% efficient.
    D. 40% efficient.

The correct answer is C.

17. Here is your next question: The output voltage of a separately excited ac generator (running at a constant speed) is controlled by:

    A. Input voltage frequency.
    B. The load.

C. Primary voltage.
D. Field current.

The correct answer is D.

Selective fading

18. What is a selective fading effect?
    A. A fading effect caused by small changes in beam heading at the receiving station.
    B. A fading effect caused by phase differences between radio wave components of the same transmission, as experienced at the receiving station.
    C. A fading effect caused by large changes in the height of the ionosphere as experienced at the receiving station.
    D. A fading effect caused by time differences between the receiving and transmitting stations.

The correct answer is B.

19. Here is your next question: What is the propagation effect called when phase differences between radio wave components of the same transmission are experienced at the recovery station?
    A. Faraday rotation.
    B. Diversity reception.
    C. Selective fading.
    D. Phase shift.

The correct answer is C. This is a good example of a question that is also a definition.

20. Here is your next question: What is the major cause of selective fading?
    A. Small changes in beam heading at the receiving station.
    B. Large changes in the height of the ionosphere, as experienced at the receiving station.
    C. Time differences between the receiving and transmitting stations.
    D. Phase differences between radio wave components of the same transmission, as experienced at the receiving station.

The correct answer is D.

21. Here is your next question: Which emission modes suffer the most from selective fading?
    A. CW and SSB.
    B. FM and double sideband AM.
    C. SSB and image.
    D. SSTV and CW.

The correct answer is B.

22. Here is your next question: How does the bandwidth of the transmitted signal affect selective fading?

    A. It is more pronounced at wide bandwidths.
    B. It is more pronounced at narrow bandwidths.
    C. It is equally pronounced at both narrow and wide bandwidths.
    D. The receiver bandwidth determines the selective fading effect.

The correct answer is A.

Absorption wavemeter

23. An absorption wave meter is useful in measuring:

    A. Field strength.
    B. Output frequencies to conform with FCC tolerance.
    C. Standing wave frequencies.
    D. The resonant frequency of LC tank circuit.

The correct answer is D.

Operational amplifiers

24. What is an operational amplifier?

    A. A high-gain, direct-coupled differential amplifier whose characteristics are determined by components external to the amplifier unit.
    B. A high-gain, direct-coupled audio amplifier whose characteristics are determined by components external to the amplifier unit.
    C. An amplifier used to increase the average output of frequency modulated signals.
    D. A program subroutine that calculates the gain of an RF amplifier.

The correct answer is A.

25. Here is your next question: What would be the characteristics of the ideal op-amp?

    A. Zero input impedance, infinite output impedance, infinite gain, flat frequency response.
    B. Infinite input impedance, zero output impedance, infinite gain, flat frequency response.
    C. Zero input impedance, zero output impedance, infinite gain, flat frequency response.
    D. Infinite input impedance, infinite output impedance, infinite gain, flat frequency response.

The correct answer is B.

26. Here is your next question: What determines the gain of a closed-loop op-amp circuit?

A. The external feedback network.

B. The collector-to-base capacitance of the pnp stage.

C. The power supply voltage.

D. The pnp collector load.

The correct answer is A.

27. Here is your next question: What is meant by the term op-amp offset voltage?

A. The output voltage of the op-amp minus its input voltage.

B. The difference between the output voltage of the op-amp and the input voltage required in the following stage.

C. The potential between the amplifier-input terminals of the op-amp in a closed-loop condition.

D. The potential between the amplifier-input terminals of the op-amp in an open-loop condition.

The correct answer is C.

28. Here is your next question: What is the input impedance of a theoretically ideal op-amp?

A. 100 Ω.

B. 1000 Ω.

C. Very low.

D. Very high.

The correct answer is D.

29. Here is your next question: What is the output impedance of a theoretically ideal op-amp?

A. Very low.

B. Very high.

C. 100 Ω.

D. 1000 Ω.

The correct answer is A.

30. Here is your next question: What determines the gain and frequency characteristics of an op-amp RC active filter?

A. Values of capacitances and resistances built into the op-amp.

B. Values of capacitances and resistances external to the op-amp.

C. Voltage and frequency of dc input to the op-amp power supply.

D. Regulated dc voltage output from the op-amp power supply.

The correct answer is B.

31. Here is your next question: What are the principle uses of an op-amp RC active filter?

    A. Op-amp circuits are used as high-pass filters to block RF1 at the input receivers.

    B. Op-amp circuits are used as low-pass filters between transmitters and transmission lines.

    C. Op-amp circuits are used as filters for smoothing power-supply output.

    D. Op-amp circuits are used as audio filters for receivers.

The correct answer is D.

32. Here is your next question: What type of capacitors should be used in an op-amp RC active filter circuit?

    A. Electrolytic.

    B. Disc ceramic.

    C. Polystyrene.

    D. Paper dielectric.

The correct answer is C.

33. Here is your next question: How can unwanted ringing and audio instability be prevented in a multisection op-amp RC audio filter circuit?

    A. Restrict both gain and Q.

    B. Restrict gain, but increase Q.

    C. Restrict Q, but increase gain.

    D. Increase both gain and Q.

The correct answer is A.

34. Here is your next question: Where should an op-amp RC active audio filter be placed in a receiver?

    A. In the IF strip, immediately before the detector.

    B. In the audio circuitry immediately before the speaker or phone jack.

    C. Between the balanced modulator and frequency multiplier.

    D. In the low-level audio stages.

The correct answer is D.

35. Here is your next question: What parameter must be selected when designing an audio filter using an op-amp?

    A. Bandpass characteristics.

    B. Desired current gain.

    C. Temperature coefficient.

    D. Output-offset overshoot.

The correct answer is A.

36. Here is your next question: What is an inverting op-amp circuit?
    A. An operational amplifier circuit connected such that the input and output signals are 180 degrees out of phase.
    B. An operational amplifier circuit connected such that the input and output signals are in phase.
    C. An operational amplifier circuit connected such that the input and output signals are 90 degrees out of phase.
    D. An operational amplifier circuit connected such that the input impedance is held at zero, while the output impedance is high.

The correct answer is A.

37. Here is your next question: What is a noninverting op-amp circuit?
    A. An operational amplifier circuit connected such that the input and output signals are 180 degrees out of phase.
    B. An operational amplifier circuit connected such that the input and output signals are in phase.
    C. An operational amplifier circuit connected such that the input and output signals are 90 degrees out of phase.
    D. An operational amplifier circuit connected such that the input impedance is held at zero while the output impedance is high.

The correct answer is B.

38. Here is your next question: How does the gain of a theoretically ideal operational amplifier vary with frequency?
    A. The gain increases linearly with increasing frequency.
    B. The gain decreases linearly with increasing frequency.
    C. The gain decreases logarithmically with increasing frequency.
    D. The gain does not vary with frequency.

The correct answer is D.

Phase-locked loops (PLL)

39. What is a phase-locked loop circuit?
    A. An electronic servo loop consisting of a ratio detector, reactance modulator and voltage-controlled oscillator.
    B. An electronic circuit also known as a monostable multivibrator.
    C. An electronic circuit consisting of a precision push-pull amplifier with a differential input.
    D. An electronic servo loop consisting of a phase detector, a low-pass filter and voltage-controlled oscillator.

The correct answer is D. An amplifier can be included in the feedback loop.

40. Here is your next question: What functions are performed by a phase-locked loop?

    A. Wideband AF and RF power amplification.
    B. Comparison of two digital input signals, digital pulse counter.
    C. Photovoltaic conversion, optical coupling.
    D. Frequency synthesis, FM demodulation.

The correct answer is D.

41. Here is your next question: A circuit compares the output from a voltage-controlled oscillator and a frequency standard. The difference between the two frequencies produces an error voltage that changes the voltage-controlled oscillator frequency. What is the name of the circuit?

    A. A doubly balanced mixer
    B. A phase-locked loop.
    C. A differential voltage amplifier.
    D. A variable frequency oscillator.

The correct answer is B. If the frequencies of the VCO and the frequency standard are the same a correction voltage is not needed. When that happens the PLL is in a locked condition.

Circuits

42. What is the voltage drop across R1? (Please refer to Fig. 1-1.):

    A. 9 V.
    B. 7 V.
    C. 5 V.
    D. 3 V.

The correct answer is C.

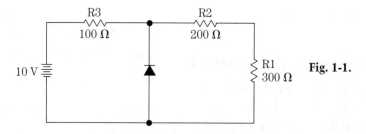

Fig. 1-1.

43. Here is your next question: What is the voltage drop across R1? (Please refer to Fig. 1-2.):

    A. 1.2 V.
    B. 2.4 V.
    C. 3.7 V.
    D. 9 V.

The correct answer is D.

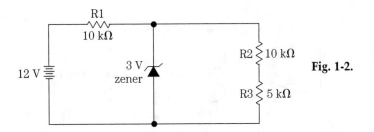

Fig. 1-2.

44. Here is your next question: In a properly operating marine transmitter, if the power supply bleeder resistor opens:

    A. Short circuit of supply voltage due to overload.
    B. Regulation would decrease.
    C. Next stage would fail due to short circuit.
    D. Filter capacitors might short from voltage surge.

The correct answer is D.

45. Here is your next question: Which of the following can occur that would least affect this circuit? (Please refer to Fig. 1-3.):

    A. C1 shorts.
    B. C1 opens.
    C. C3 shorts.
    D. C18 opens.

The correct answer is D.

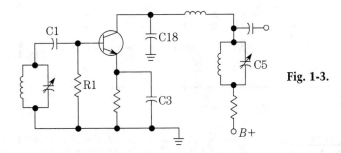

Fig. 1-3.

46. Here is your next question: When S1 is closed, lights L1 and L2 go on. What is the condition of both lamps when both S1 and S2 are closed? (Please refer to Fig. 1-4.):

    A. Both lamps stay on.
    B. L1 turns off; L2 stays on.
    C. Both lamps turn off.
    D. L1 stays on; L2 turns off.

The correct answer is D.

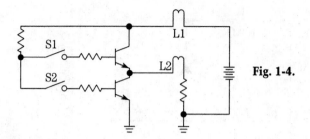

Fig. 1-4.

47. Here is your next question: If S1 is closed, both lamps light, what happens when S1 and S2 are closed? (Please refer to Fig. 1-4.):

    A. L1 and L2 are off.
    B. L1 is on and L2 is flashing.
    C. L1 is off and L2 is on.
    D. L1 is on and L2 is off.

The correct answer is D.

48. Here is your next question: How can you correct the defect, if any, in this voltage doubler circuit? (Please refer to Fig. 1-5.):

    A. Omit C1.
    B. Reverse polarity signs.
    C. Ground X.
    D. Reverse polarity on C1.

The correct answer is B.

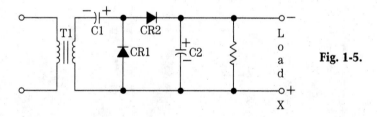

Fig. 1-5.

49. Here is your next question: What change is needed in order to correct the grounded emitter amplifier shown? (Please refer to Fig. 1-6.):

    A. No change is necessary.
    B. Polarities of emitter-base battery should be reversed.
    C. Polarities of collector-base battery should be reversed.
    D. Point A should be replaced with a low value capacitor.

The correct answer is A.

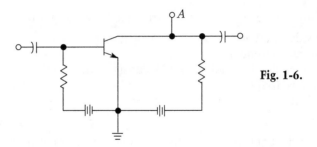

Fig. 1-6.

Frequency counters

50. What is a frequency standard?

    A. A well-known (standard) frequency used for transmitting certain messages.
    B. A device used to produce a highly accurate reference frequency.
    C. A device for accurately measuring frequency to within 1 Hz.
    D. A device used to generate wide-band random frequencies.

The correct answer is B.

51. Here is your next question: What is a frequency-marker generator?

    A. A device used to produce a highly accurate reference frequency.
    B. A sweep generator.
    C. A broadband white-noise generator.
    D. A device used to generate wide-band random frequencies.

The correct answer is A. The markers appear on the oscilloscope screen as blips or bright spots. They are used to show reference frequency points on a sweep generator oscilloscope display.

52. Here is your next question: How is a frequency-marker generator used?

    A. In conjunction with a grid-dip meter.
    B. To provide reference points on a receiver dial.
    C. As the basic frequency element of a transmitter.
    D. To directly measure wavelength.

The correct answer is B. This is another use of a marker generator.

53. Here is your next question: What is a frequency counter?

    A. A frequency measuring device.
    B. A frequency marker generator.
    C. A device that determines whether or not a given frequency is in use before automatic transmissions are made.
    D. A broadband white-noise generator.

The correct answer is A.

54. Here is your next question: How is a frequency counter used?

    A. To provide reference points on an analog receiver dial.
    B. To generate a frequency standard.
    C. To measure the deviation in an FM transmitter.
    D. To measure frequency.

The correct answer is D.

55. Here is your next question: What factors limit the accuracy, frequency response, and stability of a frequency counter?

    A. Number of digits in the readout, speed of the logic and time base stability.
    B. Time base accuracy, speed of the logic and time base stability.
    C. Time base accuracy, temperature coefficient of the logic and time base stability.
    D. Number of digits in the readout, external frequency reference and temperature coefficient of the logic.

The correct answer is B.

56. Here is your next question: How can the accuracy of a frequency counter be improved?

    A. By using slower digital logic.
    B. By improving the accuracy of the frequency response.
    C. By increasing the accuracy of the time base.
    D. By using faster digital logic.

The correct answer is C.

57. Here is your next question: What does the accuracy of a frequency counter depend on?

    A. The internal crystal reference.
    B. A voltage-regulated power supply with an unvarying output.
    C. Accuracy of the ac input frequency to the power supply.
    D. Proper balancing of the power-supply diodes.

The correct answer is A. This answer does not conflict with the answer to number 55. They mean of the choices given here.

Dip meters

58. What is a dip meter?

    A. A field strength meter.
    B. An SWR meter.
    C. A variable LC oscillator with metered feedback current.
    D. A marker generator.

The correct answer is C.

59. Here is your next question: Why is a dip meter used by many technicians?

    A. It can measure signal strength accurately.
    B. It can measure frequency accurately.
    C. It can measure transmitter output power accurately.
    D. It can give an indication of the resonant frequency of a circuit.

The correct answer is D.

60. Here is your next question: How does a dip meter function?

    A. Reflected waves at a specific frequency desensitize the detector coil.
    B. Power coupled from an oscillator causes a decrease in metered current.
    C. Power from a transmitter cancels feedback current.
    D. Harmonics of the oscillator cause an increase in resonant circuit Q.

The correct answer is B.

61. Here is your next question: What two ways could a dip meter be used in a radio station?

    A. To measure resonant frequency of antenna traps and to measure percentage of modulation.
    B. To measure antenna resonance and to measure percentage of modulation.
    C. To measure antenna resonance and to measure antenna impedance.
    D. To measure resonant frequency of antenna traps and to measure a tuned circuit resonant frequency.

The correct answer is D.

62. Here is your next question: What types of coupling occurs between a dip meter and a tuned circuit being checked?

    A. Resistive and inductive.
    B. Inductive and capacitive.
    C. Resistive and capacitive.
    D. Strong field.

The correct answer is B.

63. Here is your next question: How tight should the dip meter be coupled with the tuned circuit being checked?

    A. As loosely as possible, for best accuracy.
    B. As tightly as possible, for best accuracy.
    C. First loose, then tight, for best accuracy.
    D. With a soldered jumper wire between the meter and the circuit to be checked, for best accuracy.

The correct answer is A.

64. Here is your next question: What happens in a dip meter when it is too tightly coupled with the tuned circuit being checked?

    A. Harmonics are generated.
    B. A less accurate reading results.
    C. Cross modulation occurs.
    D. Intermodulation distortion occurs

The correct answer is B.

Oscilloscopes

65. What factors limit the accuracy, frequency response, and stability of an oscilloscope?

    A. Sweep oscillator quality and deflection amplifier bandwidth.
    B. Tube face voltage increments and deflection amplifier voltage.
    C. Sweep oscillator quality and tube face voltage increments.
    D. Deflection amplifier output impedance and tube face frequency increments.

The correct answer is A.

66. Here is your next question: How can the frequency response of an oscilloscope be improved?

    A. By using a triggered sweep and a crystal oscillator as the time base.
    B. By using a crystal oscillator as the time base and increasing the vertical sweep rate.
    C. By increasing the vertical sweep rate and the horizontal amplifier frequency response.
    D. By increasing the horizontal sweep rate and the vertical amplifier frequency response.

The correct answer is D.

Meters

67. What factors limit the accuracy, frequency response, and stability of a D'Arsonval movement type meter?

    A. Calibration, coil impedance, and meter size.
    B. Calibration, series resistance, and electromagnet current.
    C. Coil impedance, electromagnet voltage, and movement mass.
    D. Calibration, mechanical tolerance, and coil impedance.

The correct answer is D.

Spectrum analyzer

68. How does a spectrum analyzer differ from a conventional time-domain oscilloscope?

A. The oscilloscope is used to display electrical signals while the spectrum analyzer is used to measure ionospheric reflection.
B. The oscilloscope is used to display electrical signals in the frequency domain while the spectrum analyzer is used to display electrical signals in the time domain.
C. The oscilloscope is used to display electrical signals in the time domain while the spectrum analyzer is used to display electrical signals in the frequency domain.
D. The oscilloscope is used for displaying audio frequencies and the spectrum analyzer is used for displaying radio frequencies.

The correct answer is C.

69. Here is your next question: What does the horizontal axis of a spectrum analyzer display?

A. Amplitude.
B. Voltage.
C. Resonance.
D. Frequency.

The correct answer is D.

70. Here is your next question: What does the vertical axis of a spectrum analyzer display?

A. Amplitude.
B. Duration.
C. Frequency.
D. Time.

The correct answer is A.

71. Here is your next question: What test instrument can be used to display spurious signals in the output of a radio transmitter?

A. A spectrum analyzer.
B. A wattmeter.
C. A logic analyzer.
D. A time-domain reflectometer.

The correct answer is A.

72. Here is your next question: What test instrument is used to display intermodulation distortion products from an SSB transmitter?

A. A wattmeter.
B. A spectrum analyzer.
C. A logic analyzer.
D. A time-domain reflectometer

The correct answer is B.

Logic probe

73. What advantage does a logic probe have over a voltmeter for monitoring logic states in a circuit?

   A. A logic probe has fewer leads to connect to a circuit than a voltmeter.
   B. A logic probe can be used to test analog and digital circuits.
   C. A logic probe can be powered by commercial ac lines.
   D. A logic probe is smaller and shows a simplified readout.

The correct answer is D.

74. Here is your next question: What piece of test equipment can be used to directly indicate high and low logic states?

   A. A galvanometer.
   B. An electroscope.
   C. A logic probe.
   D. A Wheatstone bridge.

The correct answer is C.

75. Here is your next question: What is a logic probe used to indicate?

   A. A short-circuit fault in a digital-logic circuit.
   B. An open-circuit failure in a digital-logic circuit.
   C. A high-impedance ground loop.
   D. High and low logic states in a digital-logic circuit.

The correct answer is D. Some logic probes also indicate a pulse.

76. Here is your next question: What piece of test equipment besides an oscilloscope can be used to indicate pulse conditions in a digital-logic circuit?

   A. A logic probe.
   B. A galvanometer.
   C. An electroscope.
   D. A Wheatstone bridge.

The correct answer is A.

Marker Generator

77. What is a crystal-controlled marker generator?

   A. A low-stability oscillator that "sweeps" through a band of frequencies.
   B. An oscillator often used in aircraft to determine the craft's location relative to the inner and outer markers at airports.

    C. A high-stability oscillator whose output frequency and amplitude can be varied over a wide range.

    D. A high-stability oscillator that generates a series of reference signals at known frequency intervals.

The correct answer is D.

78. Here is your next question: What additional circuitry is required in a 100-kHz crystal-controlled marker generator to provide markers at 50 and 25 kHz?

    A. An emitter-follower.
    B. Two frequency multipliers.
    C. Two flip-flops.
    D. A voltage divider.

The correct answer is C.

Prescaler

79. What is the purpose of a prescaler circuit?

    A. It converts the output of a JK flip-flop to that of an RS flip-flop.
    B. It multiplies an HF signal so that low-frequency counter can display the operating frequency.
    C. It prevents oscillation in a low-frequency counter circuit.
    D. It divides an HF signal so that a low-frequency counter can display the operating frequency.

The correct answer is D.

Power supplies and filters

80. What is a linear electronic voltage regulator?

    A. A regulator that has a ramp voltage as its output.
    B. A regulator in which the pass transistor switches from the "off" state to the "on" state.
    C. A regulator in which the control device is switched on or off, with the duty cycle proportional to the line or load conditions.
    D. A regulator in which the conduction of a control element is varied in direct proportion to the line voltage or load current.

The correct answer is D. That is the answer supplied by the FCC.

81. Here is your next question: What is a switching electronic voltage regulator?
    A. A regulator in which the conduction of a control element is varied in direct proportion to the line voltage or load current.
    B. A regulator that provides more than one output voltage.
    C. A regulator in which the control device is switched on or off, with the duty cycle proportional to the line or load conditions.
    D. A regulator that gives a ramp voltage at its output.

The correct answer is C.

82. Here is your next question: What device is usually used as a stable reference voltage in a linear voltage regulator?
    A. A Zener diode.
    B. A tunnel diode.
    C. An SCR.
    D. A varactor diode.

The correct answer is A.

83. Here is your next question: What type of linear regulator is used in applications requiring efficient utilization of the primary power source?
    A. A constant-current source.
    B. A series regulator.
    C. A shunt regulator.
    D. A shunt current source.

The correct answer is B.

84. Here is your next question: What type of linear voltage regulator is used in applications where the load on the unregulated voltage source must be kept constant?
    A. A constant-current source.
    B. A series regulator.
    C. A shunt current source.
    D. A shunt regulator.

The correct answer is D.

85. Here is your next question: To obtain the best temperature stability, what should be the operating voltage of the reference diode in a linear voltage regulator?
    A. Approximately 2.0 V.
    B. Approximately 3.0 V.
    C. Approximately 6.0 V.
    D. Approximately 10.0 V.

The correct answer is C. That is the answer given by the FCC.

86. Here is your next question: What is the meaning of the term *remote sensing*, with regard to a linear voltage regulator?

    A. The feedback connection to the error amplifier is made directly to the load.
    B. Sensing is accomplished by wireless inductive loops.
    C. The load connection is made outside the feedback loop.
    D. The error amplifier compares the input voltage to the reference voltage.

The correct answer is A. That is the answer supplied by the FCC.

87. Here is your next question: What is a three-terminal regulator?

    A. A regulator that supplies three voltages with variable current.
    B. A regulator that supplies three voltages at a constant current.
    C. A regulator containing three error amplifiers and sensing transistors.
    D. A regulator containing a voltage reference, error amplifier, sensing resistors and transistors, and a pass element.

The correct answer is D.

88. Here is your next question: What are the important characteristics of a three-terminal regulator?

    A. Maximum and minimum input voltage, minimum output current, and voltage.
    B. Maximum and minimum input voltage, maximum output current, and voltage.
    C. Maximum and minimum input voltage, minimum output current, and maximum output voltage.
    D. Maximum and minimum input voltage, minimum output voltage, and maximum output current.

The correct answer is B.

89. Here is your next question: What is an L-network?

    A. A network consisting entirely of four inductors.
    B. A network consisting of an inductor and a capacitor.
    C. A network used to generate a leading phase angle.
    D. A network used to generate a lagging phase angle.

The correct answer is B.

90. Here is your next question: What is a pi-network?

    A. A network consisting entirely of four inductors or four capacitors.
    B. A power incidence network.
    C. An antenna matching network that is isolated from ground.
    D. A network consisting of one inductor and two capacitors or two inductors and one capacitor.

The correct answer is D.

91. Here is your next question: What is a pi-L-network?

    A. A phase-inverter load network.
    B. A network consisting of two inductors and two capacitors.
    C. A network with only three discrete parts.
    D. A matching network in which all components are isolated from ground.

The correct answer is B.

92. Here is your next question: Does the L-, pi-, or pi-L-network provide the greatest harmonic suppression?

    A. L-network.
    B. Pi-network.
    C. Inverse L-network.
    D. Pi-L-network.

The correct answer is D.

93. Here is your next question: What are the three most commonly used networks to accomplish a match between an amplifying device and a transmission line?

    A. M-network, pi-network and T-network.
    B. T-network, M-network and Q-network.
    C. L-network, pi-network and pi-L-network.
    D. L-network, M-network and C-network.

The correct answer is C.

94. Here is your next question: How are networks able to transform one impedance to another?

    A. Resistances in the networks substitute for resistances in the load.
    B. The matching network introduces negative resistance to cancel the resistive part of an impedance.
    C. The matching network introduces transconductance to cancel the reactive part of an impedance.
    D. The matching network can cancel the reactive part of an impedance and change the value of the resistive part of an impedance.

The correct answer is D.

95. Here is your next question: Which type of network offers the greater transformation ratio?

    A. L-network.
    B. Pi-network.
    C. Constant-K.
    D. Constant-M.

The correct answer is B.

96. Here is your next question: Why is the L-network of limited utility in impedance matching?

    A. It matches a small impedance range.
    B. It has limited power handling capabilities.
    C. It is thermally unstable.
    D. It is prone to self resonance.

The correct answer is A.

97. Here is your next question: What is an advantage of using a pi-L-network instead of a pi-network for impedance matching between the final amplifier of a vacuum tube type transmitter and a multiband antenna?

    A. Greater transformation range.
    B. Higher efficiency.
    C. Lower losses.
    D. Greater harmonic suppression.

The correct answer is D.

98. Here is your next question: Which type of network provides the greatest harmonic suppression?

    A. L-network.
    B. Pi-network.
    C. Pi-L-network.
    D. Inverse-pi network.

The correct answer is C.

99. Here is your next question: What are the three general groupings of filters?

    A. High-pass, low-pass, and bandpass.
    B. Inductive, capacitive, and resistive.
    C. Audio, radio, and capacitive.
    D. Hartley, Colpitts, and Pierce.

The correct answer is A.

100. Here is your next question: What is a constant-k filter?

    A. A filter that uses Boltzmann's constant.
    B. A filter whose velocity factor is constant over a wide range of frequencies.
    C. A filter whose product of the series- and shunt-element impedances is a constant for all frequencies.
    D. A filter whose input impedance varies widely over the design bandwidth.

The correct answer is C.

101. Here is your next question: What is an advantage of a constant-k filter?
  A. It has high attenuation for signals on frequencies far removed from the passband.
  B. It can match impedances over a wide range of frequencies.
  C. It uses elliptic functions.
  D. The ratio of the cutoff frequency to the trap frequency can be varied.

The correct answer is A.

102. Here is your next question: What is an m-derived filter?
  A. A filter whose input impedance varies widely over the design bandwidth.
  B. A filter whose product of the series- and shunt-element impedances is a constant for all frequencies.
  C. A filter whose schematic shape is the letter "M."
  D. A filter that uses a trap to attenuate undesired frequencies too near cutoff for a constant-k filter.

The correct answer is D. That is the answer supplied by the FCC.

103. Here is your next question: What are the distinguishing features of a Butterworth filter?
  A. A filter whose product of the series- and shunt-element impedances is a constant for all frequencies.
  B. It only requires capacitors.
  C. It has a maximally flat response over its passband.
  D. It requires only inductors.

The correct answer is C.

104. Here is your next question: What are the distinguishing features of a Chebyshev filter?
  A. It has a maximally flat response over its passband.
  B. It allows ripple in the passband.
  C. It only requires inductors.
  D. A filter whose product of the series- and shunt-element impedances is a constant for all frequencies.

The correct answer is B.

Teletype (RTTY)

As with other Q&A questions supplied by the FCC, you will find that the questions and answers in this section define the various parameters in RTTY. Even if you have not studied RTTY, you will learn the necessary basics by studying this section. You can do the same thing with every section in these books.

105. What is one advantage of using the ASCII code, with its larger character set, instead of the Baudot code?

    A. ASCII includes built-in error-correction features.

    B. ASCII characters contain fewer information bits than Baudot characters.

    C. It is possible to transmit upper- and lowercase text.

    D. The larger character set allows store-and-forward control characters to be added to a message.

The correct answer is C.

106. Here is your next question: What is the duration of a 45-baud Baudot RTTY data pulse?

    A. 11 ms.

    B. 40 ms.

    C. 31 ms.

    D. 22 ms.

The correct answer is D.

107. Here is your next question: What is the duration of a 45-baud Baudot RTTY start pulse?

    A. 11 ms.

    B. 22 ms.

    C. 31 ms.

    D. 40 ms.

The correct answer is B.

108. Here is your next question: What is the duration of a 45-baud Baudot RTTY stop pulse?

    A. 11 ms.

    B. 18 ms.

    C. 31 ms.

    D. 40 ms.

The correct answer is C.

109. Here is your next question: What is the necessary bandwidth of a 170-Hz shift, 45-baud Baudot emission F1B transmission?

    A. 45 Hz.

    B. 249 Hz.

    C. 442 Hz.

    D. 600 Hz.

The correct answer is B.

110. Here is your next question: What is the necessary bandwidth of a 170-Hz shift, 45-baud Baudot emission J2B transmission?

    A. 45 Hz.
    B. 249 Hz.
    C. 442 Hz.
    D. 600 Hz.

The correct answer is B.

111. Here is your next question: What is the necessary bandwidth of a 170-Hz shift, 74-baud Baudot emission F1B transmission?

    A. 250 Hz.
    B. 278 Hz.
    C. 442 Hz.
    D. 600 Hz.

The correct answer is B.

112. Here is your next question: What is the necessary bandwidth of a 170-Hz shift, 74-baud Baudot emission J2B transmission?

    A. 250 Hz.
    B. 278 Hz.
    C. 442 Hz.
    D. 600 Hz.

The correct answer is B.

113. Here is your next question: What is the necessary bandwidth of a 1000-Hz shift, 1200-baud ASCII emission F1D transmission?

    A. 1000 Hz.
    B. 1200 Hz.
    C. 440 Hz.
    D. 2400 Hz.

The correct answer is D.

114. Here is your next question: What is the necessary bandwidth of a 4800-Hz frequency shift, 9600-baud ASCII emission F1D transmission?

    A. 15.36 kHz.
    B. 9.6 kHz.
    C. 4.8 kHz.
    D. 5.76 kHz.

The correct answer is A.

115. Here is your next question: What is the necessary bandwidth of a 4800-Hz frequency shift, 9600-baud ASCII emission J2D transmission?

    A. 15.36 kHz.
    B. 9.6 kHz.
    C. 4.8 kHz.
    D. 9.76 kHz.

The correct answer is A.

116. Here is your next question: What is the necessary bandwidth of a 170-Hz shift, 110-baud ASCII emission F1B transmission?

    A. 304 Hz.
    B. 314 Hz.
    C. 608 Hz.
    D. 628 Hz.

The correct answer is B.

117. Here is your next question: What is the necessary bandwidth of a 170-Hz shift, 110-baud ASCII emission J2B transmission?

    A. 304 Hz.
    B. 314 Hz.
    C. 608 Hz.
    D. 628 Hz.

The correct answer is B.

118. Here is your next question: What is the necessary bandwidth of a 170-Hz shift, 300-baud ASCII emission F1D transmission?

    A. 0 Hz.
    B. 0.3 kHz.
    C. 0.5 kHz.
    D. 1.0 kHz.

The correct answer is C.

119. Here is your next question: What is the necessary bandwidth of a 170-Hz shift, 300-baud ASCII emission J2D transmission?

    A. 0 Hz.
    B. 0.3 kHz.
    C. 0.5 kHz.
    D. 1.0 kHz.

The correct answer is C.

Satellites

120. What is an ascending pass for a satellite?

    A. A pass from west to east.

    B. A pass from east to west.
    C. A pass from south to north.
    D. A pass from north to south.

The correct answer is C.

121. Here is your next question: What is a descending pass for a satellite?

    A. A pass from north to south.
    B. A pass from west to east.
    C. A pass from east to west.
    D. A pass from south to north.

The correct answer is A.

122. Here is your next question: What is the period of a satellite?

    A. An orbital arc that extends from 60 degrees west longitude to 145 degrees west longitude.
    B. A point on an orbit where satellite height is minimum.
    C. The amount of time it takes a satellite to complete an orbit.
    D. The amount of time it takes a satellite to perigee to apogee.

The correct answer is C.

123. Here is your next question: What is a linear transponder?

    A. A repeater that passes only linear or binary signals.
    B. A device that receives and transmits signals of any mode in a certain pass-band.
    C. An amplifier for SSB transmissions.
    D. A device used to change an FM emission to an AM emission.

The correct answer is B.

124. Here is your next question: What are the two basic types of linear transponders used in satellites?

    A. Inverting and noninverting.
    B. Geostationary and elliptical.
    C. Phase 2 and phase 3.
    D. Amplitude modulated and frequency modulated.

The correct answer is A.

125. Here is your next question: Why does the downlink frequency appear to vary by several kHz during a low-earth orbit satellite pass?

    A. The distance between the satellite and ground station is changing, causing the Kepler effect.
    B. The distance between the satellite and ground station is changing, causing the Bernoulli effect.

C. The distance between the satellite and ground station is changing, causing the Boyle's law effect.

D. The distance between the satellite and ground station is changing, causing the Doppler effect.

The correct answer is D.

126. Here is your next question: Why does the received signal from a satellite stabilized by a computer-pulsed electromagnet exhibit a fairly rapid pulsed fading effect?

A. Because the satellite is rotating.

B. Because of ionospheric absorption.

C. Because of the satellite's low orbital attitude.

D. Because of the Doppler effect.

The correct answer is A.

127. Here is your next question: What type of antenna can be used to minimize the effects of spin modulation and Faraday rotation?

A. A nonpolarized antenna.

B. A circularly polarized antenna.

C. An isotropic antenna.

D. A log-periodic dipole array.

The correct answer is B.

Facsimile

128. What is facsimile?

A. The transmission of characters by radio teletype that forms a picture when printed.

B. The transmission of still pictures by slow-scan television.

C. The transmission of video by television.

D. The transmission of printed pictures for permanent display on paper.

The correct answer is D.

129. Here is your next question: What is the modern standard scan rate for a facsimile picture transmitted by a radio station?

A. The modern standard is 240 lines per minute.

B. The modern standard is 50 lines per minute.

C. The modern standard is 150 lines per second.

D. The modern standard is 60 lines per second.

The correct answer is A.

130. Here is your next question: What is the approximate transmission time for a facsimile picture transmitted by a radio station?

A. Approximately 6 minutes per frame at 240 lines per minute.
B. Approximately 3.3 minutes per frame at 240 lines per minute.
C. Approximately 6 minutes per frame at 240 lines per minute.
D. ⅟₆₀ second per frame at 240 lines per minute.

The correct answer is B.

131. Here is your next question: What is the term for the transmission of printed pictures by radio for the purpose of permanent display?

A. Television.
B. Facsimile.
C. Xerography.
D. ACSSB.

The correct answer is B.

132. Here is your next question: In facsimile, how are variations of picture brightness and darkness converted into voltage variations?

A. With an LED.
B. With a Hall-effect transistor.
C. With a photodetector.
D. With an optoisolator.

The correct answer is C.

2
CHAPTER

Radar review

Note: The requirements for certification tests are not reviewed in this chapter. The review material given here is for FCC licensing only. If you are interested in certification in Radar obtain review information from the certifying agency.

An *acronym* is a word formed from the first (or, first few) letters of several words. Radar is an acronym formed from the words *RA*dio *D*irection *A*nd *R*anging.

The words describe the major functions of radar. It uses radio waves to detect the presence of aircraft, ships, land masses, etc. In addition to detecting their presence, it is also used to determine their direction and distance.

The speed of radio waves is about the same as the speed of light:

> 186,000 statute miles per second, or
> 162,000 nautical miles per second, or
> 300,000,000 meters per second

In its simplest form, radar sends a pulse of electromagnetic energy, which is a short burst of radio wave in this case, toward an object. The object is referred to as a *target* (see Fig. 2-1). The time it takes the pulse to go to the target and bounce back to the source is measured. Using the simple equation:

$$distance = rate \times time$$

the distance can be obtained when the rate and time are known.
Distance in:

> *statute miles* = 186,000 miles/sec × *seconds*, or
> *nautical miles* = 162,000 nautical miles/sec × *seconds*, or
> *meters* = 300,000,000 meters/second × *seconds*

Convenient units of measurement for radio wave speeds can be derived from the above speeds:

> *pulse speed* = 0.186 miles per microsecond time for radio pulse to travel 1 mile
>
> = 10.75 µs the pulse travels 32 yards/µs

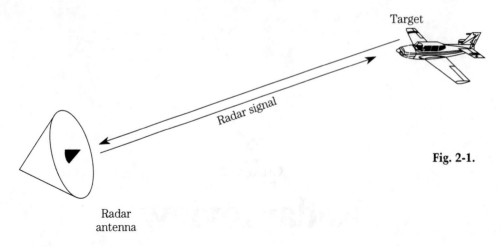

Fig. 2-1.

Remember that speed is the rate of change of distance. Velocity is speed in a given direction.

Examples
speed: 400 miles per hour
velocity: 400 miles per hour north

The time it takes for the pulse to leave the radar antenna and return is the *round trip time.* That time has to be divided by 2 to get the time it takes to get to the target. It is the time the pulse takes to get to the target that is needed to determine the target distance.

Example
If it takes 163 microseconds for a radar pulse to reach a target, how far is the target from the radar station?

Solution
Take the reciprocal of 10.75 microseconds per mile to get 0.093 miles per microsecond. Then: (0.093 miles/microsecond) × (163 microseconds) = approximately 15.2 miles.

A directional antenna makes it possible to determine the direction of the target from the radar station. Two very directional antennas are shown in Fig. 2-2. The Yagi type is lighter, but the parabolic type has better overall characteristics for locating the target. The directional antenna is turned until the return pulse from the target produces the strongest signal in the antenna. Other methods are also used.

If a land-based radar system is used to locate an aircraft, three things must be determined:

1. The distance to the target (called the *slant distance*),
2. The azimuth, or, horizontal direction to the target (also called the *bearing*), and,
3. The elevation or vertical angle to the target. The method of determining the distance to the target has already been covered. The azimuth and elevation

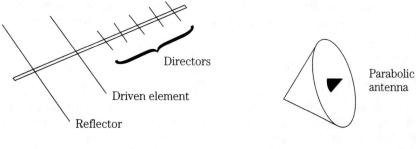

Directors

Driven element

Reflector

Yagi antenna

Parabolic antenna

Fig. 2-2.

are usually determined by sensing devices that show what direction the antenna is pointing for the strongest signal.

There are devices in the antenna that provide signals for determining its position. A servomechanism is an example. In one application, a servomechanism rotates the deflection coils of a CRT in step with a rotating antenna. The result is a display called a *plan position indicator (PPI)*.

If the slant distance is known, and the angle the slant distance makes with a horizontal line is known, additional useful information can be obtained. A computer can be used to make the necessary calculations. Figure 2-3 shows the definitions and necessary equations.

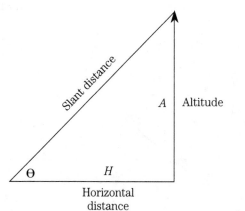

Slant distance

A | Altitude

Θ | H

Horizontal distance

Fig. 2-3.

Altitude = Slant distance

Horizontal distance = Slant distance × cos. Θ

Altitude = Slant distance × sin Θ

The speed of the target can be obtained mathematically by using information obtained from the radar measurements. The basic equation relating angular velocity and radius is used.

linear velocity = angular velocity × radius

The angular velocity is the rate at which the antenna must be turned to follow the target; and, the radius is the slant distance. Again, a computer can perform the mathematics.

Displays

A land-based or ship-based radar can make at least three measurements on an aircraft: distance, bearing, and elevation. However, the radar display is presented on a cathode ray tube that gives a two-dimension pattern. Although some very clever innovations have been tried for displaying all three measurements, the usual pattern displays the target with only two of the three measurements on a flat screen. Figure 2-4 shows the more popular examples of displays.

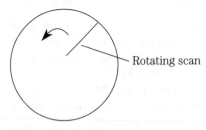

Plan position indicator

Fig. 2-4.

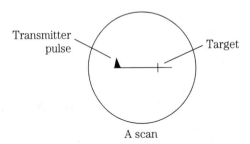

A scan

The signal

Two different types of signals are used in radar systems. The pulsed signal just described is the more popular one. A continuous sine-wave signal (CW) is often used when speed is the measurement desired. It utilizes a Doppler effect to locate a moving target.

One way to understand the Doppler effect is to imagine a train coming toward you. If its horn is blowing you hear a certain frequency. As the train passes the frequency of the horn decreases.

In a somewhat similar manner, if you direct a sine-wave carrier toward a target that is approaching the return signal will have a phase shift. The amount of phase shift will increase if the speed of the target is increased; and, there will be a decrease in frequency if the speed of the target decreases. To summarize, the speed of the aircraft can be determined by measuring the amount of phase shift in the reflected RF carrier.

The pulsed radar signal

The pulsed radar signal is obtained by turning the RF carrier on and off. The on-off signal is obtained from a section called the *synchronizer* or *timer*.

The number of pulses generated per second is called the *pulse repetition frequency (PRF)*. That frequency is important because there must be enough time between pulses to permit the pulse to reach the target and return.

The width of the radar pulse determines the minimum range of the radar system. During the time that the transmitter is on (delivering a pulse), there can be no information obtained from returned pulse energy.

Example

The width of a radar pulse is one microsecond. What is the minimum range of the radar system?

Solution

In 1 microsecond, the radar pulse can travel 164 yards, so there can be no target closer than 164 yards.

As with any kind of transmitter, the power in the transmitted signal is a factor in determining how far the signal will go. Because the transmitted signal is pulsed, the output power can be very high during pulse transmission, but the average power might be low (Fig. 2-5).

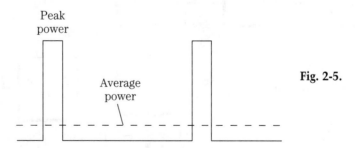

Peak power

Average power

Fig. 2-5.

It is important to know the average power of the transmitted signal. The duty cycle (the on time divided by the total time for one cycle) for a pulsed radar signal is defined in the following ways:

$$duty\ cycle = Pulse\ width/pulse\ repetition\ frequency$$

$$duty\ cycle = on\ time/time\ for\ one\ complete\ cycle$$

$$duty\ cycle = average\ power/peak\ power$$

So:

$$average\ power = peak\ power \times duty\ cycle$$

Radar frequencies Higher RF frequencies are desirable for radar operation because transmission on the surface is limited to line of sight, that is, the distance to the horizon. However, not every radar system operates at line-of-sight frequencies. For example, high-power, low-frequency radars are now in operation that go well beyond the line of sight. They are used for weather forecasting.

Here are some important line-of-sight radar frequencies:

2900–3100 MHz	S band	wavelength = 10.3–9.7 cm
5470–5650 MHz	X band	wavelength = 5.5–5.3 cm
9000–9500 MHz	X band	wavelength = 3.22–3.15 cm
Airborne radar: 10,000 MHz		wavelength = 3 cm
Airborne radar: 30,000 MHz		wavelength = 1 cm

Here is a convenient equation for calculating wavelength in centimeters:

$$Wavelength \text{ (cm)} = 30,000/frequency \text{ (MHz)}$$

A radar transceiver

Figure 2-6 shows the block diagram of a typical pulsed radar transceiver.

The *transmitter* has a high-power pulse generator. Magnetrons are frequently used as the transmitter output device. The signal from the magnetron, or other transmitter output device, is controlled by a modulator. The *receiver* is a crystal input device. The received signal is converted to a lower IF frequency at the crystal stage. The crystal is actually a semiconductor diode.

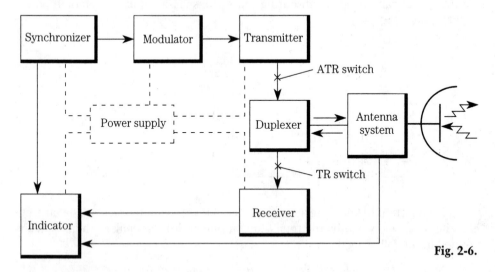

Fig. 2-6.

The *TR (transmit/receive) switch* protects the crystal and other receiver components when radar transmission is taking place. The *ATR (anti-transmit/receive) switch* prevents the received pulse from being wasted in the transmitter section. The *duplexer* makes it possible to transmit and receive using the same antenna. The *indicator* is very much like an oscilloscope. It puts the display on the screen of the cathode ray tube (CRT). The *synchronizer* (also known as the *timer*) supplies timing pulses to the modulator in the transmitter. Those pulses determine the pulse repetition frequency and the pulse duration.

The synchronizer also supplies a pulse to the indicator circuitry. That pulse starts the trace on the CRT the instant the transmitter pulse stops (Fig. 2-7). All of the dc operating voltages are produced by the power supply.

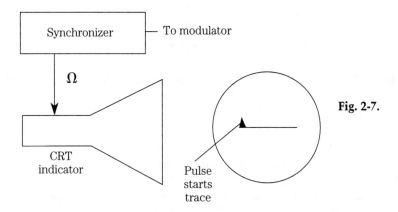

Fig. 2-7.

TR and ATR switches

Before discussing the operation of the TR switches, it is important to review the characteristics of waveguide planes. Electromagnetic energy moving down a waveguide, or in space, consists of two fields: electric and magnetic. When a waveguide is designed to favor the electric field, it is said to be an E-plane device. A waveguide that favors the magnetic field is an H-plane device.

Figure 2-8 shows an example of a TR (transmit/receive) switch. It is used for two purposes in a radar waveguide:

1. When used as a TR switch it prevents a high-power radar pulse from destroying the input crystal and other components in the receiver during transmission, and,

2. When used in a ATR switch it prevents the return pulse from being dissipated in the transmitter.

There is a gas between the electrodes in the tube. That gas is ionized by a transmitter pulse. Ionized gas in a TR switch acts like a short circuit across a waveguide.

A *keep-alive voltage* maintains the gas in the TR tube at the point where it is almost ionized. A strong transmitter pulse completes the ionization.

The paths in the TR and ATR sections have different lengths. In the quarter-wave lengths, the ionized TR switch acts as a short circuit and reflects an open circuit at the input. When the TR switch is not ionized, the return signal passes through the waveguide without obstruction.

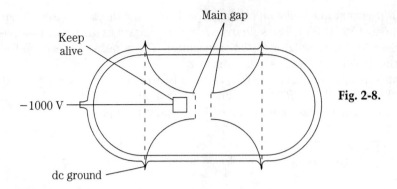

Fig. 2-8.

In the ATR section, the longer path presents an open circuit when the TR switch is not fired. That prevents the weak received signal from entering the transmitter section when the system is in the receive mode.

The hybrid ring duplexer

In a pulsed radar system, a transmitter is either sending a high-energy pulse or receiving a weak signal. It does not do both things at the same time. The job of the duplexer is to switch between transmission and reception. Switching can be done automatically in a radar system by using a *hybrid ring duplexer* like the one in Fig. 2-9.

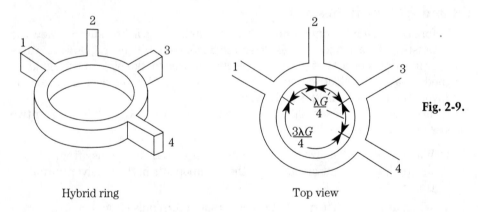

Hybrid ring   Top view

Fig. 2-9.

The characteristics of quarter-wave and half-wave shorted-end and open-end waveguides are the same as for transmission lines. For example, a shorted quarter-wave transmission line is an open circuit at the entrance. Therefore, when the TR tubes fire, they act as a short circuit at a quarter wavelength and that reflects an open circuit at the entrance.

During reception the TR switches are off because there is no ionization. The received signal enters the branch going to the receiver. However, the ATR branch now offers a one-half wavelength shorted stub to the signal. It acts like an open circuit at its entrance. That is what keeps the return signal out of the transmitter section.

Special high-frequency tubes

In any test for radar licensing, you can expect questions on magnetrons, klystrons, and traveling wave tubes. The programmed section in this chapter will give you a review on those subjects. If you cannot answer the review questions, you should do some additional study before attempting to sit for a radar license test.

Programmed review

Start with Block 1. Pick the answer that you believe is correct. Go to the next block and check your answer. All answers are in italics. There is only one choice for each block.

Block 1

Here is your first question: What is the power limitation of associated ship stations operating under the authority of a ship station license?

A. The power level authorized to the parent ship station.
B. Associated vessels are prohibited from operating under the authority granted to another station licensee.
C. The minimum power necessary to complete the radio communications.
D. Power is limited to one watt.

Block 2

The correct answer is D.

Here is your next question: Licensed radiotelephone operators are not required on board ships for:

A. voluntarily equipped ship stations on domestic voyages operating on VHF channels.
B. ship radar, provided that the equipment is nontunable, with a pulse-type magnetron, and it can be operated by means of exclusively external controls.
C. installation of a VHF transmitter in a ship station where the work is performed by or under the immediate supervision of the licensee of the ship station.
D. any of the above.

Block 3

The correct answer is D.

Here is your next question: In the International Phonetic Alphabet, the letters E, M, and S are represented by the words:

A. Echo, Michigan, Sonar.
B. Equator, Mike, Sonar.
C. Echo, Mike, Sierra.
D. Element, Mister, Scooter.

Block 4

The correct answer is C.

Here is your next question: If the elapsed time for a radar echo is 62 microseconds, what is the distance in nautical miles to the object?

A. 5.
B. 87.
C. 37.
D. 11.5.

Block 5

The correct answer is A.

Here is your next question: What is the wavelength of a signal at 500 MHz?

A. 0.062 cm.
B. 6 m.
C. 60 cm.
D. 60 m.

Block 6

The correct answer is C.

Here is your next question: The radar range in nautical miles to an object can be found by measuring the elapsed time during a radar pulse and dividing this quantity by:

A. 0.87 s.
B. 1.15 µs.
C. 12.36 µs.
D. 1.73 µs.

Block 7

The correct answer is C.

Here is your next question: A high standing wave ratio on a transmission line can be caused by:

A. excessive modulation.
B. an increase in output power.
C. detuned antenna coupling.
D. poor B+ regulation.

Block 8

The correct answer is C.

Here is your next question: The best insulation at UHF is:

A. black rubber.
B. bakelite.

C. paper.
D. mica.

Block 9

The correct answer is D.

Here is your next question: The super-high frequency (SHF) band is:

A. 3000 to 30,000 MHz.
B. above 300,000 MHz.
C. 300 to 3000 MHz.
D. 30,000 to 300,000 MHz.

Block 10

The correct answer is A.

Here is your next question: Magnetron oscillators are used for:

A. generating SHF signals.
B. multiplexing.
C. generating rich harmonics.
D. GM demodulation.

Block 11

The correct answer is A.

Here is your next question: Velocity of radio wave propagation in free space does what?

A. Varies in speed.
B. Is constant.
C. Is same as speed as light.
D. Both B and C above.

Block 12

The correct answer is D.

Here is your next question: Waveguides are:

A. a hollow tube that carries RF.
B. solid conductor of RF.
C. coaxial cables.
D. copper wire.

Block 13

The correct answer is A.

Here is your next question: A circulator:

A. cools dc motors during heavy loads.
B. allows two or more antennas to feed one transmitter.

C. allows one antenna to feed two separate microwave transmitters and receivers at the same time.

D. insulates UHF frequencies on transmission lines.

Block 14

The correct answer is C.

Here is your next question: What type of antenna system allows you to receive and transmit at the same time?

A. Simplex.
B. Duplex.
C. Multiplex.
D. Digital diplex.

Block 15

The correct answer is B.

Here is your next question: What is a microwave device that allows RF energy to pass through in one direction with very little loss, but absorbs RF power in the opposite direction?

A. Circulator.
B. Wave trap.
C. Multiplex.
D. Isolator.

Block 16

The correct answer is D.

Here is your next question: What is azimuth on a radar antenna?

A. Diameter.
B. Degrees elevation.
C. Degrees horizontal.
D. Impedance.

Block 17

The correct answer is C.

Here is your next question: What is the skin effect?

A. The phenomenon where RF current flows in a thinner layer of the conductor, close to the surface, as frequency increases.
B. The phenomenon where RF current flows in a thinner layer of the conductor, close to the surface, as frequency decreases.
C. The phenomenon where thermal effects on the surface of the conductor increase the impedance.
D. The phenomenon where thermal effects on the surface on the conductor decrease the impedance.

Block 18

The correct answer is A.

Here is your next question: What is the dielectric constant for air?

A. Approximately 1.
B. Approximately 2.
C. Approximately 4.
D. Approximately 0.

Block 19

The correct answer is A.

Here is your next question: How can a neon lamp be used to check for the presence of RF?

A. A neon lamp will go out in the presence of RF.
B. A neon lamp will change color in the presence of RF.
C. A neon lamp will light on in the presence of very low frequency RF.
D. A neon lamp will light in the presence of RF.

Block 20

The correct answer is D.

Here is your next question: Signal energy is coupled into a traveling-wave tube at the:

A. collector end of the helix.
B. anode end of the helix.
C. cathode end of the helix.
D. focusing coils.

Block 21

The correct answer is C.

Here is your next question: The permanent magnetic field that surrounds a traveling-wave tube (TWT) is intended to:

A. provide a means of coupling.
B. prevent the electron beam from spreading.
C. prevent oscillations.
D. prevent spurious oscillations.

Block 22

The correct answer is B.

Here is your next question: Electromagnetic coils encase a traveling-wave tube to:

A. provide a means of coupling energy.
B. prevent the electron beam from spreading.
C. prevent oscillation.
D. prevent spurious oscillation.

Block 23

The correct answer is A.

Here is your next question: What is the flywheel effect?

A. The continued motion of a radio wave through space when the transmitter is turned off.
B. The back and forth oscillation of electrons in an LC circuit.
C. The use of a capacitor in a power supply to filter rectified ac.
D. The transmission of a radio signal to a distant station by several hops through the ionosphere.

Block 24

The correct answer is B.

Here is your next question: What is an astable multivibrator?

A. A circuit that alternates between two stable states.
B. A circuit that alternates between a stable state and an unstable state.
C. A circuit set to block either a 0 pulse or a 1 pulse and pass the other.
D. A circuit that alternates between two unstable states.

Block 25

The correct answer is D.

Here is your next question: What are electromagnetic waves?

A. Alternating currents in the core of an electromagnet.
B. A wave consisting of two electric fields at right angles to each other.
C. A wave consisting of an electric field and a magnetic field at right angles to each other.
D. A wave consisting of two magnetic fields at right angles to each other.

Block 26

The correct answer is C.

Here is your next question: How does the gain of a parabolic dish antenna change when the operating frequency is doubled?

A. Gain does not change.
B. Gain is multiplied by 0.707.
C. Gain increases 6 dB.
D. Gain increases 3 dB.

Block 27

The correct answer is C.

Here is your next question: What ferrite rod device prevents the formation of reflected waves on a waveguide transmission line?

A. Reflector.

B. Isolator.
C. Wave-trap.
D. SWR refractor.

Block 28

The correct answer is B.

Here is your next question: Frequencies most affected by knife-edge refraction are:

A. low and medium frequencies.
B. high frequencies.
C. very high and ultra high frequencies.
D. 100 kHz to 3.0 MHz.

Block 29

The correct answer is C.

Here is your next question: What allows microwaves to pass in only one direction?

A. RF emitter.
B. Ferrite isolator.
C. Capacitor.
D. Varactor-triac.

Block 30

The correct answer is B.

Here is your next question: Waveguides are:

A. used exclusively in high-frequency power supplies.
B. ceramic couplers attached to antenna terminals.
C. high-pass filters used at low radio frequencies.
D. hollow metal conductors used to carry high-frequency current.

Block 31

The correct answer is D.

Here is your next question: Waveguide construction:

A. should not use silver plating.
B. should not use copper.
C. should have short vertical runs.
D. should not have long horizontal runs.

Block 32

The correct answer is D.

Here is your next question: To couple energy into and out of a waveguide:

A. use wide copper sheeting.
B. use an LC circuit.

C. use capacitive coupling.
D. use a thin piece of wire as an antenna.

Block 33

The correct answer is D.

Here is your next question: An isolator:

A. acts as a buffer between a microwave oscillator coupled to a waveguide.
B. acts as a buffer to protect a microwave oscillator from variations in line load changes.
C. shields UHF circuits from RF transfer.
D. both A and B.

Block 34

The correct answer is D.

Here is your next question: Waveguides are:

A. used exclusively in high-frequency power supplies.
B. ceramic couplers attached to antenna terminals.
C. high-pass filters used at low radio frequencies.
D. hollow metal conductors used to carry high-frequency current.

Block 35

The correct answer is D.

Here is your next question: Conductance occurs in a waveguide:

A. by interelectron delay.
B. through electrostatic field reluctance.
C. in the same manner as a transmission line.
D. through electromagnetic and electrostatic fields in the walls of the waveguide.

Block 36

The correct answer is D.

Here is your next question: How is the radar pulse repetition frequency (PRF) related to maximum range?

A. Not related.
B. One of the determining factors.
C. The only determining factor.

Block 37

The correct answer is B. Another determining factor is the strength of the radar signal. Signal strength always influences range with any kind of transmitted signal. When a pulse is transmitted there must be time for that pulse to get to the target and back before the next pulse is sent. Suppose the PRF is 500 pulses per second and the pulse width is 5 microseconds. The time for one cycle is easily determined by the equation:

$$T = \frac{1}{f} = \frac{1}{500} = 0.002 \text{ s} = 2000 \text{ μs}$$

The pattern is illustrated in Fig. 2-10. For most calculations, you can disregard the 5-μs pulse, but it is included here to make a complete solution.

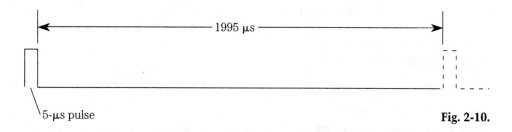

5-μs pulse

Fig. 2-10.

The "listening time" is 1995 μs. That is the maximum time for the pulse round trip. One-half of that time is the maximum time allowed for the pulse to get to the target.

$$\frac{T}{2} = \frac{1995}{2} = 997.5 \text{ μs}$$

$$distance = rate \times time$$

$$= 186{,}000 \frac{\text{miles}}{\text{seconds}} \times 0.0009975 \text{ seconds}$$

$$= 185 \text{ miles}$$

That is the slant range.

Here is your next question: In the magnetron illustration of Fig. 2-11, what are the holes used for?

A. Ventilation.
B. Determining frequency of oscillation.

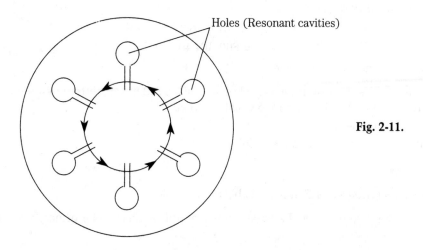

Holes (Resonant cavities)

Fig. 2-11.

Block 38

The correct answer is B. The "holes" are actually resonant cavities; along with the slots, they act like L-C circuits. The slots represent C and the holes represent L.

Here is your next question: Is the following equation correct?

$$Wavelength\ (in\ cm) = \frac{30,000}{Frequency\ (in\ MHz)}$$

A. The equation is correct.
B. The equation is not correct.

Block 39

The correct answer is B.

Here is your next question: What is the name of the circuit that controls the magnetron output?

A. Inverter.
B. Converter.
C. Modulator.
D. Impeller.

Block 40

The correct answer is C.

Here is your next question: If the slant range to a target is 800 yards and the target elevation is 40 degrees, what is the altitude of the target above earth?

A. About 3133 ft.
B. About 1543 ft.
C. About 900 ft.
D. About 86.6 ft.

Block 41

The correct answer is B.

$$altitude = slant\ range \times SIN\ A$$
$$= 800\ SIN\ 40°$$

Where: *A* is the elevation angle.

Here is your next question: A small hole is discovered in the bottom of a horizontal waveguide run. Which of the following is correct?

A. Replace the waveguide.
B. It is probably there for a purpose!

Block 42

The correct answer is B. It is probably a drain hole.

Here is your next question: Radar systems that utilize the Doppler principle are used for:

A. targets that are not moving.
B. targets that are moving.

Block 43

The correct answer is B.

Here is your next question: What is the name of the circuit that establishes the PRF?

A. Magic T.
B. Duplexer.
C. Synchronizer.

Block 44

The correct answer is C.

Here is your next question: Magnetron strapping assures:

A. safety of personnel.
B. operation at a single (desired) frequency.

Block 45

The correct answer is B.

Here is your next question: What is the name of the voltage that maintains a TR tube at near ionization?

A. Keep alive.
B. Reflex.
C. 1/R.

Block 46

The correct answer is A.

Here is your next question: If a radar receiver mixer crystal is defective, the result might be:

A. loss of target on the display.
B. weak or fuzzy target on the display.
C. poor receiver sensitivity indicated.
D. by an echo box test.
E. all of the above.

Block 47

The correct answer is E.

Here is your next question: If the indication is that a magnetron might be weak, several sharp blows on the magnet can improve the magnetron operation. This statement is:

A. correct.
B. not correct.

Block 48

The correct answer is B.

Here is your next question: An echo box provides:

A. sound reverberation to enhance noise.
B. a target for radar testing.
C. a method of testing for multipath distortion.
D. none of the above.

Block 49

The correct answer is B.

Here is your next question: A defective magnetron can cause:

A. weak or fuzzy target displays.
B. loss of range markers.

Block 50

The correct answer is A.

Here is your next question: A weak magnetron magnet can cause:

A. low magnetron current.
B. excessive magnetron current.

Block 51

The correct answer is B.

Here is your next question: Range markers on a PPI display are:

A. bright spots that make circles on the display.
B. horizontal lines in the display.

Block 52

The correct answer is A.

Here is your next question: It takes 220 microseconds for a certain radar pulse to reach the target and return. How far is the target? (Give the answer in miles.)

A. About 20.5 miles.
B. About 33.3 miles.
C. About 40.92 miles.

Block 53

The correct answer is A. Remember, 220 μs is the round trip time.

$$Distance = \frac{Rate \times Time}{2}$$

$$= \frac{186{,}000 \times 220 \times 10^{-6}}{2} = 20.46 \text{ miles}$$

Here is your next question: The ability of a radar system to distinguish between two close-spaced targets at the same range is called:

A. range distinction.
B. range delineation.
C. range derivative.
D. none of the above.

Block 54

The correct answer is D. It is called bearing resolution.

Here is your next question: In what section of a radar receiver would you expect to find a discriminator?

A. The pulse forming network.
B. The AGC.
C. The AFC.
D. The high-voltage supply.

Block 55

The correct answer is C.

Here is your next question: For what reason is a sensitivity time control (STC) used to reduce the radar receiver for a period immediately after the transmission of a radar pulse?

A. To prevent distortion of the radar pulse.
B. To reduce or eliminate returns from nearby structures.
C. To prevent interference from local TV.

Block 56

The correct answer is B.

Here is your next question: In a radar system, average power divided by peak power equals:

A. apparent power.
B. true power.
C. duty cycle.
D. PRF.

Block 57

The correct answer is C.

Here is your next question: A pulse delay network can be made with:

A. a neon lamp in parallel with a resistor.
B. a length of transmission line.

Block 58

The correct answer is B.

Here is your next question: What is a Klystron used for in a radar system?

A. A local oscillator.
B. A phase inverter.

Block 59

The correct answer is A.

Here is your next question: What advantage, if any, does a yagi antenna have in radar over a parabolic antenna?

A. It is easier to clean.
B. It is easier to match to a 300-Ω transmission line.
C. It does not require impedance matching.
D. It is lighter, so it is easier to turn.

Block 60

The correct answer is D.

Here is your next question: At radar microwave frequencies, a waveguide has:

A. higher loss than a transmission line.
B. lower loss than a transmission line.

Block 61

The correct answer is B.

Here is your next question: A magnetron is a:

A. diode.
B. triode.
C. tetrode.
D. pentode.

Block 62

The correct answer is A.

Here is your next question: The quarter-wave slot in a choke joint reflects a short circuit at the joint. This statement is:

A. correct.
B. not correct.

Block 63

The correct answer is A. Note: The quarter-wave slot acts like a half-wave shorted stub because it is a quarter wavelength in and a quarter wavelength out. Therefore, it acts like a short circuit at the input.

Here is your next question: An ATR tube is located in the:

A. transmitter waveguide path.
B. receiver waveguide path.

Block 64

The correct answer is B.

Here is your next question: Why are rectangular waveguides usually preferred over circular waveguides?

A. They are cheaper.
B. They do not require choke joints.
C. The rectangular waveguides are less likely to produce changes in polarization at junction and bends.
D. It is easier to get a signal out of a rectangular waveguide.

Block 65

The correct answer is C.

3
CHAPTER

Amateur radio

The questions and answers in this chapter are supplied by the FCC as study material for the amateur technician no-code license exam. The question-and-answer method of study has worked very well for this type of exam. Answers are at the end of this chapter.

You should consider chapters 1 through 7 as part of your training for taking the technician no-code license exam. Also, you will find the following definitions and comments to be a useful review.

Definitions

ANSI American National Standards institute

ARES American Radio Emergency Service. An ARRL-organized emergency service.

ARRL American Radio Relay League. A well-known amateur radio organization.

BAUD The reciprocal of the shortest time used for sending a unit of data. If the shortest unit of data takes 0.8 ms, the maximum rate of data transmission is:

$$Baud = \frac{1}{0.8 \times 10^{-3}} = 1250$$

That is the maximum signal rate. *Note:* The term *baud rate* has no meaning because baud is already a rate.

Bits per second This is a rate at which information is sent in the form of data.

Broadcasting Transmission of news, music, and other entertainment.

Control point The location where the station operation is controlled. The control point might not be located near the transmitter.

Courtesy tone An audio signal that tells when a transmission on a repeater is ended.

Digipeater A repeater used for digital communications.

EIA *Electronic Industries Association.* An organization that generates standards.

FSK *Frequency Shift Keying* (see chapter 5)

Ham Amateur radio operator

Ionosphere A layer of ions above the earth. It is used for reflecting radio waves back to earth, and thereby making long-distance communications possible. Bending of radio waves is less likely to occur at higher frequencies such as VHF (very-high frequencies).

Layers The ionosphere is divided into individual layers. Each layer affects long-distance communications.

Line impedance Most amateur radio transmission lines have an impedance of 50 ohms.

LUF Lowest Usable Frequency

MODEM MODulator/DEModulator. The term *modem* is used to describe a method of sending information on telephone lines by using computers at both ends.

MUF Maximum Usable Frequency

Overdeviation What occurs when an FM signal goes beyond the limits set by the FCC. It can be caused by yelling into the microphones.

Packet A method of sub-dividing data for ease of communications. It is especially useful for transfer of data between computers.

Packet radio A method of data communications in which data is divided into packets.

PDM Pulse Duration Modulation

Q signals Signals that are used to reduce the time for communications.

Examples of *Q* signals

QRP Reduced Power. This *Q* designation is also used to designate very-low power communications.

QST Call to members of ARRL

QTH Location

QRM Noise or interference

QSL Acknowledgement of a transmission.

Other examples are given in the FCC questions in this chapter. For a more complete list of *Q* signals consult *The ARRL Handbook for Radio Amateurs*.

RACES Radio Amateur Civil Emergency Services

Reactance Modulator A circuit that produces phase modulation. It changes the phase of the carrier by using the audio signal to vary circuit inductance or capacitance.

Repeater A transceiver that automatically re-transmits amateur signals. It makes communications possible between low-power amateur transceivers.

Repeater-Timeout The amount of time allotted by a repeater for transmission of a signal is limited. "Timeout" ends a transmission that exceeds the time limit.

RTTY Radio TeleTYpe

RTTY-Frequency Shift At frequencies above 50 MHz (megahertz) the FCC does not limit RTTY or packet radio bandwidth.

Secondary User In some cases amateurs are allowed to share frequency bands with other services. The service that has been assigned the band has priority, and is called the primary user.

Sky Waves Radio waves that are reflected by the ionosphere.

SOS An emergency call used on CW (continuous wave) transmission.

SWR Standing wave radio.

Yagi A highly directional antenna made with a driven element, a reflector and one or more directors.

Practice questions for the amateur no-code technician license

The questions in this section have been provided by the FCC. Some questions are repeated throughout the chapter to give you re-enforcement of your review. Related subjects are covered in the companion book titled *The TAB Sourcebook For Communications Licensing and Certification Examinations.* The answers are on pages 102, 103, and 104.

1. What does horizontal wave polarization mean?
    A. The magnetic lines of force of a radio wave are parallel to the earth's surface.
    B. The electric lines of force of a radio wave are parallel to the earth's surface.
    C. The electric lines of force of a radio wave are perpendicular to the earth's surface.
    D. The electric and magnetic lines of force of a radio wave are perpendicular to the earth's surface.

2. How is a Yagi antenna constructed?
    A. Two or more straight, parallel elements are fixed in line with each other.
    B. Two or more square or circular loops are fixed in line with each other.
    C. Two or more square or circular loops are stacked inside each other.
    D. A straight element is fixed in the center of three or more elements which angle toward the ground.

3. What type of beam antenna uses two or more straight elements arranged in line with each other?
    A. A delta loop antenna.
    B. A quad antenna.
    C. A Yagi antenna.
    D. A Zepp antenna.

4. What electromagnetic-wave polarization does a Yagi antenna have when its elements are parallel to the earth's surface?
    A. Circular.
    B. Helical.
    C. Horizontal.
    D. Vertical.

5. If you install a 6-meter Yagi antenna on a tower 150 feet from your transmitter, which of the following feedlines is best?

    A. RG-213.
    B. RG-58.
    C. RG-59.
    D. RG-174.

6. What is a parasitic beam antenna?

    A. An antenna where some elements obtain their radio energy by induction or radiation from a driven element.
    B. An antenna where wave traps are used to magnetically couple the elements.
    C. An antenna where all elements are driven by direct connection to the feed-line.
    D. An antenna where the driven element obtains its radio energy by induction or radiation from director elements.

7. What are the "parasitic elements" of a Yagi antenna?

    A. The driven element and any reflectors.
    B. The director and the driven element.
    C. Only the reflectors (if any).
    D. Any directors or any reflectors.

8. What is a delta loop antenna?

    A. A type of cubical quad antenna, except with triangular elements, rather than square.
    B. A large copper ring or wire loop, used in direction finding.
    C. An antenna system made of three vertical antennas, arranged in a triangular shape.
    D. An antenna made from several triangular coils of wire on an insulating form.

9. Why might a dummy antenna get warm when in use?

    A. Because it changes RF energy into heat.
    B. Because it absorbs static electricity.
    C. Because it stores radio waves.
    D. Because it stores electric current.

10. What is a cubical quad antenna?

    A. Four straight, parallel elements in line with each other, each approximately ½-electrical wavelength long.
    B. Two or more parallel four-sided wire loops, each approximately one-electrical wavelength long.
    C. A vertical conductor ¼-electrical wavelength high, fed at the bottom.
    D. A center-fed wire ¼-electrical wavelength long.

11. If a magnetic-base whip antenna is placed on the roof of a car, in what direction does it send out radio energy?

    A. It goes out equally well in all horizontal directions.

B. Most of it goes in one direction.

C. Most of it goes equally in two opposite directions.

D. Most of it is aimed high into the air.

12. What is the term for the average power supplied to an antenna transmission line during one RF cycle at the crest of the modulation envelope?

A. Peak transmitter power.

B. Peak output power.

C. Average radio-frequency power.

D. Peak envelope power.

13. What electromagnetic-wave polarization does a half-wavelength antenna have when it is perpendicular to the earth's surface?

A. Circular.

B. Horizontal.

C. Parabolical.

D. Vertical.

14. How many directly driven elements do most beam antennas have?

A. None.

B. One.

C. Two.

D. Three.

15. Why should you regularly clean, tighten, and resolder all antenna connectors?

A. To help keep their resistance at a minimum.

B. To keep them looking nice.

C. To keep them from getting stuck in place.

D. To increase their capacitance.

16. If your antenna feedline gets hot when you are transmitting, what might this mean?

A. You should transmit using less power.

B. The conductors in the feedline are not installed very well.

C. The feedline is too long.

D. The SWR may be too high, or the feedline loss might be high.

17. What is a directional antenna?

A. An antenna which sends and receives radio energy equally well in all directions.

B. An antenna that cannot send and receive radio energy by skywave or skip propagation.

C. An antenna which sends and receives radio energy mainly in one direction.

D. An antenna which sends and receives radio energy equally well in two opposite directions.

18. What type of antenna is made when a magnetic-base whip antenna is placed on the roof of a car?

    A. A cubical quad.
    B. A ground plane.
    C. A Yagi.
    D. A delta loop.

19. For RF safety, what is the best thing to do with your transmitting antennas?

    A. Use vertical polarization.
    B. Use horizontal polarization.
    C. Mount the antennas where no one can come near them.
    D. Mount the antenna close to the ground.

20. Which type of antenna would be a good choice as part of a portable HF amateur station that could be set up in case of an emergency?

    A. A three-element quad.
    B. A three-element Yagi.
    C. A dipole.
    D. A parabolic dish.

21. What device should be connected to a transmitter's output when you are making transmitter adjustments?

    A. A dummy antenna.
    B. A receiver.
    C. A reflectometer.
    D. A multimeter.

22. Where should an RF wattmeter be connected for the most accurate readings of transmitter output power?

    A. One-half wavelength from the antenna feed point.
    B. One-half wavelength from the transmitter output.
    C. At the antenna feed point.
    D. At the transmitter output connector.

23. What is a dummy antenna?

    A. A flexible antenna usually used on hand-held transceivers.
    B. An antenna used as a reference for gain measurements.
    C. A nonradiating load for a transmitter.
    D. A nondirectional transmitting antenna.

24. Why would you use a dummy antenna?

    A. For off-the-air transmitter testing.
    B. To reduce output power.
    C. To give comparative signal reports.
    D. To allow antenna tuning without causing interference.

25. What is the main component of a dummy antenna?

    A. A wire-wound resistor.
    B. An iron-core coil.
    C. A noninductive resistor.
    D. An air-core coil.

26. What minimum rating should a dummy antenna have for use with a 100-W single-sideband phone transmitter?

    A. 100 W continuous.
    B. 141 W continuous.
    C. 175 W continuous.
    D. 200 W continuous.

27. What common connector usually joins RG-213 coaxial cable to an HF transceiver?

    A An F-type cable connector.
    B. A PL-259 connector.
    C. A banana plug connector.
    D. A binding post connector.

28. Which of these common connectors has the lowest loss at UHF?

    A. An F-type cable connector.
    B. A type-N connector.
    C. A BNC connector.
    D. A PL-259 connector.

29. What common connector usually joins a hand-held transceiver to its antenna?

    A. A BNC connector.
    B. A PL-259 connector.
    C. An F-type cable connector.
    D. A binding post connector.

30. Why should you make sure that no one can touch an open-wire feedline while you are transmitting with it?

    A. Because contact might cause a short circuit and damage the transmitter.
    B. Because contact might break the feedline.
    C. Because contact might cause spurious emissions.
    D. Because high-voltage radio energy might burn the person.

31. What device can be installed to feed a balanced antenna with an unbalanced feedline?

    A. A balun.
    B. A loading coil.
    C. A triaxial transformer.
    D. A wavetrap.

32. What is an unbalanced line?
    A. Feedline with neither conductor connected to ground.
    B. Feedline with both conductors connected to ground.
    C. Feedline with one conductor connected to ground.
    D. Feedline with both conductors connected to each other.

33. What happens to radio energy when it is sent through a poor-quality coaxial cable?
    A. It causes spurious emissions.
    B. It is returned to the transmitter's chassis ground.
    C. It is converted to heat in the cable.
    D. It causes interference to other stations near the transmitting frequency.

34. As the length of a feedline is changed, what happens to signal loss?
    A. Signal loss is the same for any length of feedline.
    B. Signal loss increases as length increases.
    C. Signal loss decreases as length increases.
    D. Signal loss is the least when the length is the same as the signal's wavelength.

35. What does forward power mean?
    A. The power traveling from the transmitter to the antenna.
    B. The power radiated from the top of an antenna system.
    C. The power produced during the positive half of an RF cycle.
    D. The power used to drive a linear amplifier.

36. What device can measure an impedance mismatch in your antenna system?
    A. A reflectometer.
    B. A wavemeter.
    C. An ammeter.
    D. A field-strength meter.

37. What does reflected power mean?
    A. The power radiated down to the ground from an antenna.
    B. The power returned to a transmitter from an antenna.
    C. The power produced during the negative half of an RF cycle.
    D. The power returned to an antenna by building and trees.

38. Where should a reflectometer be connected for best accuracy when reading the impedance match between an antenna and its feedline?
    A. At the antenna feed point.
    B. At the transmitter output connector.
    C. At the midpoint of the feedline.
    D. Anywhere along the feedline.

39. What does standing-wave ratio mean?
    A. The ratio of maximum to minimum inductances on a feedline.
    B. The ratio of maximum to minimum resistances on a feedline.
    C. The ratio of maximum to minimum impedances on a feedline.
    D. The ratio of maximum to minimum voltages on a feedline.

40. If a directional RF wattmeter read 96 W forward power and 4 W reflected power, what is the actual transmitter output power?
    A 92 W.
    B. 88 W.
    C. 80 W.
    D. 100 W.

41. As the frequency of a signal is changed, what happens to signal loss in a feedline?
    A. Signal loss is the same for any frequency.
    B. Signal loss increases with increasing frequency.
    C. Signal loss increases with decreasing frequency.
    D. Signal loss is the least when the signal's wavelength is the same as the feedline's length.

42. At what point in your station is transceiver power measured?
    A. At the power supply terminals inside the transmitter or amplifier.
    B. At the final amplifier input terminals inside the transmitter or amplifier.
    C. At the antenna terminals of the transmitter or amplifier.
    D. On the antenna itself, after the feedline.

43. Where should fuses be connected on a mobile transceiver's dc power cable?
    A. Between the red and black wires.
    B. In series with just the black wire.
    C. In series with just the red wire.
    D. In series with both the red and black wires.

44. If a 4800-Ω resistor is connected to 12 V, how much current will flow through it?
    A. 4000 A.
    B. 4000 mA.
    C. 250 mA.
    D. 2500 μA.

45. If a 12-V battery supplied 0.15 A to a circuit, what is the circuit's resistance?
    A. 80 Ω.
    B. 12 Ω.
    C. 1.8 Ω.
    D. 0.15 Ω.

46. If a 48,000-Ω resistor is connected to 120 V, how much current will flow through it?

    A. 400 A.
    B. 40 A.
    C. 25 mA.
    D. 2.5 mA.

47. If a 4800-Ω resistor is connected to 120 V, how much current will flow through it?

    A. 4 A.
    B. 25 mA.
    C. 25 A.
    D. 40 mA.

48. What does the fourth color band on a resistor indicate?

    A. The value of the resistor in ohms.
    B. The resistance tolerance in percent.
    C. The power rating in watts.
    D. The resistance material.

49. If a 12-V battery supplied 0.25 A to a circuit, what is the circuit's resistance?

    A. 0.25 Ω.
    B. 3 Ω.
    C. 12 Ω.
    D. 48 Ω.

50. How do you find a resistor's tolerance rating?

    A. By reading its Baudot code.
    B. By using Thevenin's theorem for resistors.
    C. By using a voltmeter.
    D. By reading the resistor's color code.

51. What do the first three color bands on a resistor indicate?

    A. The value of the resistor in ohms.
    B. The resistance tolerance in percent.
    C. The power rating in watts.
    D. The resistance material.

52. What does a multimeter measure?

    A. SWR and power.
    B. Resistance, capacitance and inductance.
    C. Resistance and reactance.
    D. Voltage, current and resistance.

53. Why is the retaining screw in one terminal of a wall outlet made of brass while the other one is silver colored?

    A. To prevent corrosion.
    B. To indicate correct wiring polarity.
    C. To better conduct current.
    D. To reduce skin effect.

54. Where should the white wire in a three-wire ac line cord be connected in a power supply?

    A. To the side of the power transformer's primary winding that has a fuse.
    B. To the side of the power transformer's primary winding that does not have a fuse.
    C. To the chassis.
    D. To the black wire.

55. What document would you use to see if you comply with standard electrical safety rules when building an amateur antenna?

    A. The Code of Federal Regulations.
    B. The Proceedings of the IEEE.
    C. The National Electrical Code.
    D. The ITU Radio Regulations.

56. How is an ammeter usually connected to a circuit under test?

    A. In series with the circuit.
    B. In parallel with the circuit.
    C. In quadrature with the circuit.
    D. In phase with the circuit.

57. How is a voltmeter usually connected to a circuit under test?

    A. In phase with the circuit.
    B. In quadrature with the circuit.
    C. In series with the circuit.
    D. In parallel with the circuit.

58. What happens inside a voltmeter when you switch it from a lower to a higher voltage range?

    A. Resistance is added in series with the meter.
    B. Resistance is added in parallel with the meter.
    C. Resistance is reduced in series with the meter.
    D. Resistance is reduced in parallel with the meter.

59. Where should the black (or red) wire in a three-wire ac line cord be connected in a power supply?

   A. To the fuse.
   B. To the chassis.
   C. To the green wire.
   D. To the white wire.

60. Where should the green wire in a three-wire ac line cord be connected in a power supply?

   A. To the chassis.
   B. To the "hot" side of the power switch.
   C. To the fuse.
   D. To the white wire.

61. What circuit blocks RF energy above and below a certain limit?

   A. A bandpass filter.
   B. A high-pass filter.
   C. An input filter.
   D. A low-pass filter.

62. What type of filter is used in the IF section of receivers to block energy outside a certain frequency range?

   A. A bandpass filter.
   B. A high-pass filter.
   C. An input filter.
   D. A low-pass filter.

63. What describes a capacitor?

   A. Two or more layers of silicon material with an insulating material between them.
   B. The wire used in the winding and the core material.
   C. Two or more conductive plates with an insulating material between them.
   D. Two or more insulating plates with a conductive material between them.

64. What does a capacitor do?

   A. It stores a charge electrochemically and opposes a change in current.
   B. It stores a charge electrostatically and opposes a change in voltage.
   C. It stores a charge electromagnetically and opposes a change in current.
   D. It stores a charge electromechanically and opposes a change in voltage.

65. What is a farad?

   A. The basic unit of resistance.
   B. The basic unit of capacitance.
   C. The basic unit of inductance.
   D. The basic unit of admittance.

66. What is the ability to store energy in an electric field called?

    A. Inductance.
    B. Resistance.
    C. Tolerance.
    D. Capacitance.

67. If two equal-value capacitors are connected in series, what is their total capacitance?

    A. Twice the value of one capacitor.
    B. The same as the value of either capacitor.
    C. Half the value of either capacitor.
    D. The value of one capacitor times the value of the other.

68. If two equal-value capacitors are connected in parallel, what is their total capacitance?

    A. Twice the value of one capacitor.
    B. Half the value of one capacitor.
    C. The same as the value of either capacitor.
    D. The value of one capacitor times the value of the other.

69. What is the basic unit of capacitance?

    A. The farad.
    B. The ohm.
    C. The volt.
    D. The henry.

70. As the plate area of a capacitor is increased, what happens to its capacitance?

    A. It decreases.
    B. It increases.
    C. It stays the same.
    D. It disappears.

71. What determines the capacitance of a capacitor?

    A. The material between the plates, the area of one side of one plate, the number of plates and the spacing between the plates.
    B. The material between the plates, the number of plates and the size of the wires connected to the plates.
    C. The number of plates, the spacing between the plates and whether the dielectric material is N-type or P-type.
    D. The material between the plates, the area of one plate, the number of plates and the material used for the protective coating.

72. What circuit is found in all types of receivers?

    A. An audio filter.
    B. A beat-frequency oscillator.

C. A detector.

D. An RF amplifier.

73. What precaution should you take when leaning over a power amplifier?

A. Take your shoes off.

B. Watch out for loose jewelry contacting high voltage.

C. Shield your face from the heat produced by the power supply.

D. Watch out for sharp edges which may snag your clothing.

74. Why is the limit of exposure to RF the lowest in the frequency range of 30 MHz to 300 MHz, according to the ANSI RF protection guide?

A. There are more transmitters operating in this range.

B. There are fewer transmitters operating in this range.

C. Most transmissions in this range are for a longer time.

D. The human body absorbs RF energy the most in this range.

75. What should you do if you discover someone is being burned by high voltage?

A. Run from the area so you won't be burned too.

B. Turn off the power, call for emergency help and give CPR if needed.

C. Immediately drag the person away from the high voltage.

D. Wait for a few minutes to see if the person can get away from the high voltage on their own, then try to help.

76. What is the basic unit of inductance?

A. The coulomb.

B. The farad.

C. The henry.

D. The ohm.

77. What is an inductor core?

A. The place where a coil is tapped for resonance.

B. A tight coil of wire used in a transformer.

C. Insulating material placed between the wires of a transformer.

D. The place inside an inductor where its magnetic field is concentrated.

78. What can happen if you tune a ferrite-core coil with a metal tool?

A. The metal tool can change the coil's inductance and cause you to tune the coil incorrectly.

B. The metal tool can become magnetized so much that you might not be able to remove it from the coil.

C. The metal tool can pick up enough magnetic energy to become very hot.

D. The metal tool can pick up enough magnetic energy to become a shock hazard.

79. What determines the inductance of a coil?

    A. The core material, the core diameter, the length of the coil and whether the coil is mounted horizontally or vertically.
    B. The core diameter, the number of turns of wire used to wind the coil and the type of metal used for the wire.
    C. The core material, the number of turns used to wind the core and the frequency of the current through the coil.
    D. The core material, the core diameter, the length of the coil and the number of turns of wire used to wind the coil.

80. If two equal-value inductors are connected in series, what is their total inductance?

    A. Half the value of one inductor.
    B. Twice the value of one inductor.
    C. The same as the value of either inductor.
    D. The value of one inductor times the value of the other.

81. As an iron core is inserted in a coil, what happens to the coil's inductance?

    A. It increases.
    B. It decreases.
    C. It stays the same.
    D. It disappears.

82. What is a henry?

    A. The basic unit of admittance.
    B. The basic unit of capacitance.
    C. The basic unit of inductance.
    D. The basic unit of resistance.

83. What is the ability to store energy in a magnetic field called?

    A. Admittance.
    B. Capacitance.
    C. Resistance.
    D. Inductance.

84. What does an inductor do?

    A. It stores a charge electrostatically and opposes a change in voltage.
    B. It stores a charge electrochemically and opposes a change in current.
    C. It stores a charge electromagnetically and opposes a change in current.
    D. It stores a charge electromechanically and opposes a change in voltage.

85. If two equal-value inductors are connected in parallel, what is their total inductance?

A. Half the value of one inductor.
B. Twice the value of one inductor.
C. The same as the value of either inductor.
D. The value of one inductor times the value of the other (*Note:* No inductive coupling).

86. Why do resistors sometimes get hot when in use?

A. Some electrical energy passing through them is lost as heat.
B. Their reactance makes them heat up.
C. Hotter circuit components nearby heat them up.
D. They absorb magnetic energy which makes them hot.

87. Why would a large-sized resistor be used instead of a smaller one of the same resistance?

A. For better response time.
B. For a higher current gain.
C. For greater power dissipation.
D. For less impedance in the circuit.

88. What does resistance do in an electric circuit?

A. It stores energy in a magnetic field.
B. It stores energy in an electric field.
C. It provides electrons by a chemical reaction.
D. It opposes the flow of electrons.

89. What does a variable resistor or potentiometer do?

A. Its resistance changes when ac is applied to it.
B. It transforms a variable voltage into a constant voltage.
C. Its resistance changes when its slide or contact is moved.
D. Its resistance changes when it is heated.

90. Which of the following are common resistor types?

A. Plastic and porcelain.
B. Film and wire wound.
C. Electrolytic and metal film.
D. Iron core and brass core.

91. If you know the voltage and current supplied to a circuit, what formula would you use to calculate the circuit's resistance?

A. Ohm's law.
B. Tesla's law.
C. Ampere's law.
D. Kirchoff's law.

92. How is the current in a dc circuit calculated when the voltage and resistance are known?

    A. $I = R \times E$ (current equals resistance multiplied by voltage).
    B. $I = R/E$ (current equals resistance divided by voltage).
    C. $I = E/R$ (current equals voltage divided by resistance).
    D. $I = P/E$ (current equals power divided by voltage).

93. How is the voltage in a dc circuit calculated when the current and resistance are known?

    A. $E = I/R$ (voltage equals current divided by resistance).
    B. $E = R/I$ (voltage equals resistance divided by current).
    C. $E = I \times R$ (voltage equals current multiplied by resistance).
    D. $E = P/I$ (voltage equals power divided by current).

94. How is the resistance in a dc circuit calculated when the voltage and current a known?

    A. $R = I/E$ (resistance equals current divided by current).
    B. $R = E/I$ (resistance equals voltage divided by current).
    C. $R = I \times E$ (resistance equals current multiplied by voltage).
    D. $R = P/E$ (resistance equals power divided by voltage).

95. What are the possible values of a 100-Ω resistor with a 10% tolerance?

    A. 90 to 100 Ω.
    B. 10 to 100 Ω.
    C. 90 to 110 Ω.
    D. 80 to 120 Ω.

96. Which tolerance rating would a high-quality resistor have?

    A. 0.1%.
    B. 5%.
    C. 10%.
    D. 20%.

97. Which tolerance rating would a low-quality resistor have?

    A. 0.1%.
    B. 5%.
    C. 10%.
    D. 20%.

98. Ohm's law describes the mathematical relationship between what three electrical quantities?

    A. Resistance, voltage and power.
    B. Current, resistance and power.

C. Current, voltage and power.

D. Resistance, current and voltage.

99. In Fig. 3-1, which symbol represents an inductor wound over a toroidal core?

A. Symbol A.

B. Symbol B.

C. Symbol C.

D. Symbol D.

100. In Fig. 3-1, which symbol represents an adjustable inductor?

A. Symbol A.

B. Symbol B.

C. Symbol C.

D. Symbol D.

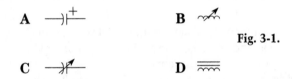

Fig. 3-1.

101. What is the condition of the ionosphere just before local sunrise?

A. Atmospheric attenuation is at a maximum.

B. The D region is above the E region.

C. The E region is above the F region.

D. Ionization is at a minimum.

102. What causes tropospheric ducting of radio waves?

A. A very low pressure area.

B. An aurora to the north.

C. Lightning between the transmitting and receiving stations.

D. A temperature inversion.

103. What is the ionosphere?

A. An area of the atmosphere where weather takes place.

B. An area of the outer atmosphere where enough ions and free electrons exist to propagate radio waves.

C. An area between two air masses of different temperature and humidity, along which radio waves can travel.

D. An ionized path in the atmosphere where lightning has struck.

104. In Fig. 3-1, which symbol represents a variable capacitor?

A. Symbol A.

B. Symbol B.

C. Symbol C.

D. Symbol D.

105. In Fig. 3-1, which symbol represents an electrolytic capacitor?

A. Symbol A.

B. Symbol B.

C. Symbol C.

D. Symbol D.

106. In Fig. 3-1, which symbol represents an iron-core inductor?

A. Symbol A.

B. Symbol B.

C. Symbol C.

D. Symbol D.

107. What document is used by almost every U.S. city as the basis for electrical safety requirements for power wiring and antennas?

A. The Code of Federal Regulations.

B. The Proceedings of the IEEE.

C. The ITU Radio Regulations.

D. The National Electrical Code.

108. How much electrical current flowing through the human body is usually fatal?

A. As little as 1/10 of an ampere.

B. Approximately 10 amperes.

C. More than 20 amperes.

D. Current flow through the human body is never fatal.

109. What is the unlabeled block in Fig. 3-2?

A. A band-pass filter.

B. A crystal oscillator.

C. A reactance modulator.

D. A rectifier modulator.

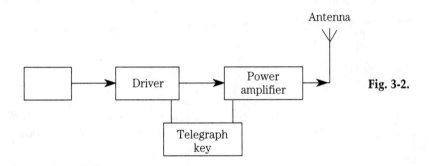

Fig. 3-2.

110. On what frequencies within the 2-meter band may image emissions be transmitted?

    A. 144.1 to 148.0 MHz only.
    B. 146.0 to 148.0 MHz only.
    C. 144.0 to 148.0 MHz only.
    D. 146.0 to 147.0 MHz only.

111. On what frequencies within the 6-meter band may phone emissions be transmitted?

    A. 50.0 to 54.0 MHz only.
    B. 50.1 to 54.0 MHz only.
    C. 51.0 to 54.0 MHz only.
    D. 52.0 to 54.0 MHz only.

112. Which operator licenses authorize privileges on 223.50 MHz?

    A. Extra, Advanced, General, Technician, Novice.
    B. Extra, Advanced, General, Technician only.
    C. Extra, Advanced, General only.
    D. Extra, Advanced only.

113. What minimum class of amateur license must you hold to operate a beacon station?

    A. Novice.
    B. Technician.
    C. General.
    D. Amateur Extra.

114. Which operator licenses authorize privileges on 146.52 MHz?

    A. Extra, Advanced, General, Technician, Novice.
    B. Extra, Advanced, General, Technician only.
    C. Extra, Advanced, General only.
    D. Extra, Advanced only.

115. Which of the following frequencies may a Technician operator who has passed a Morse code test use?

    A. 7.1 to 7.2 MHz.
    B. 14.1 to 14.2 MHz.
    C. 21.2 to 21.2 MHz.
    D. 28.1 to 29.2 MHz.

116. If you are a Novice licensee with a Certificate of Successful Completion of Examination (CSCE) for Technician privileges, how do you identify your station when transmitting on 146.34 MHz?

A. You must give your callsign, followed by any suitable word that denotes the slant mark and the identifier "KT."

B. You may not operate on 146.34 until your new license arrives.

C. No special form of identification is needed.

D. You must give your callsign and the location of the VE examination where you obtained the CSCE.

117. In addition to passing the Technician written examination (Elements 2 and 3A), what must you do before you are allowed to use amateur frequencies below 30 MHz?

A. Nothing special is needed; all Technicians may use the HF bands at any time.

B. You must notify the FCC that you intend to operate on the HF bands.

C. You must attend a class to learn about HF communications.

D. You must pass a Morse code test (either Element 1A, 1B, or 1C).

118. What is the maximum transmitter power an amateur station is allowed when used for telecommand (control) of model craft?

A. 1 mW.

B. 1 W.

C. 2 W.

D. 3 W.

119. What is the maximum transmitting power permitted an amateur station on 146.52 MHz?

A. 200 W PEP output.

B. 500 W ERP.

C. 1000 W dc input.

D. 1500 W PEP output.

120. Which operator licenses authorize privileges on 446.0 MHz?

A. Extra, Advanced, General, Technician, Novice.

B. Extra, Advanced, General, Technician only.

C. Extra, Advanced, General only.

D. Extra, Advanced only.

121. What is the usual input/output frequency separation for repeaters in the 70-centimeter band?

A. 600 kHz.

B. 1.0 MHz.

C. 1.6 MHz.

D. 5.0 MHz.

122. What electromagnetic-wave polarization does most man-made electrical noise have in the HF and VHF spectrum?

A. Horizontal.
B. Left-hand circular.
C. Right-hand circular.
D. Vertical.

123. If a repeater is causing harmful interference to another repeater and a frequency coordinator has recommended the operation of one station only, who is responsible for resolving the interference?

A. The licensee of the unrecommended repeater.
B. Both repeater licensees.
C. The licensee of the recommended repeater.
D. The frequency coordinator.

124. What is the usual input/output frequency separation for repeaters in the 1.25-meter band?

A. 600 kHz.
B. 1.0 MHz.
C. 1.6 MHz.
D. 5.0 MHz.

125. When may you send profane words from your amateur station?

A. Only when they do not cause interference to other communications.
B. Only when they are not retransmitted through a repeater.
C. Never; profane words are prohibited in amateur transmissions.
D. Any time, but there is an unwritten rule among amateurs that they should not be used on the air.

126. If you are using a frequency within a band assigned to the amateur service on a secondary basis, and a station assigned to the primary service on that band causes interference, what action should you take?

A. Notify the FCC's regional Engineer in Charge of the interference.
B. Increase your transmitter's power to overcome the interference.
C. Attempt to contact the station and request that it stop the interference.
D. Change frequencies; you may be causing harmful interference to the other station, in violation of FCC rules.

127. What information is included in the FCC declaration of a temporary state of communication emergency?

A. A list of organizations authorized to use radio communications in the affected area.
B. A list of amateur frequency bands to be used in the affected area.
C. Any special conditions and special rules to be observed during the emergency.
D. An operating schedule for authorized amateur emergency stations.

128. How must you identify messages sent during a RACES drill?

    A. As emergency messages.
    B. As amateur traffic.
    C. As official government messages.
    D. As drill or test messages.

129. What is the maximum symbol rate permitted for packet transmissions on the 10-meter band?

    A. 300 baud.
    B. 1200 baud.
    C. 19.6 kilobaud.
    D. 56 kilobaud.

130. Which of the following one-way communications may not be transmitted in the amateur service?

    A. Morse code practice.
    B. Brief transmissions to make adjustments to the station.
    C. Telecommands to model craft.
    D. Broadcasts intended for the general public.

131. What kind of payment is allowed for third-party messages sent by an amateur station?

    A. No payment of any kind is allowed.
    B. Donation of amateur equipment.
    C. Donation of equipment repairs.
    D. Any amount agreed upon in advance.

132. What rule applies if two amateur stations want to use the same frequency?

    A. The station operator with a lesser class of license must yield.
    B. The station operator with a lower power output must yield the frequency to the station with a higher power output.
    C. Both station operators have an equal right to operate on the frequency.
    D. Station operators in ITU (International Time Universal) Regions 1 and 3 must yield the frequency to stations in ITU Region 2.

133. What minimum information must be on a label affixed to a transmitter used for telecommand (control) of model craft?

    A. Station callsign.
    B. Station callsign and the station licensee's name.
    C. Station callsign and the station licensee's name and address.
    D. Station callsign and the station licensee's class of license.

134. What is the meaning of: "Your signal report is 5 by 7 . . ."?

    A. Your signal is perfectly readable and moderately strong.

  B. Your signal is perfectly readable, but weak.
  C. Your signal is readable with considerable difficulty.
  D. Your signal is perfectly readable with near pure tone.

135. What causes the maximum usable frequency to vary?

  A. The temperature of the ionosphere.
  B. The speed of the winds in the upper atmosphere.
  C. The amount of radiation received from the sun, mainly ultraviolet.
  D. The type of weather just below the ionosphere.

136. What is the maximum symbol rate permitted for RTTY or data transmissions between 28 and 50 MHz?

  A. 56 kilobaud.
  B. 19.6 kilobaud.
  C. 1200 baud.
  D. 300 baud.

137. What is the fastest code speed a repeater may use for automatic identification?

  A. 13 words per minute.
  B. 20 words per minute.
  C. 25 words per minute.
  D. There is no limitation.

138. What is the usual bandwidth of a single-sideband amateur signal?

  A. Between 2 and 3 kHz.
  B. Between 3 and 6 kHz.
  C. 2 kHz.
  D. 1 kHz.

139. In what frequency range does tropospheric ducting most often occur?

  A. VHF.
  B. HF.
  C. MF.
  D. SW.

140. What does maximum usable frequency mean?

  A. The highest frequency signal that will reach its intended destination.
  B. The lowest frequency signal that will reach its intended destination.
  C. The highest frequency signal that is most absorbed by the ionosphere.
  D. The lowest frequency signal that is most absorbed by the ionosphere.

141. In what frequency range does sky-wave propagation least often occur?

  A. LF.
  B. MF.

C. HF.

D. VHF.

142. What is meant by the upper sideband (USB)?

A. The part of a single-sideband signal that is above the carrier frequency.

B. The part of a single-sideband signal that is below the carrier frequency.

C. Any frequency above 10 MHz.

D. The carrier frequency of a single-sideband signal.

143. What is the usual bandwidth of a frequency-modulated amateur signal?

A. Less than 5 kHz.

B. Between 5 and 10 kHz.

C. Between 10 and 20 kHz.

D. Greater than 20 kHz.

144. What happens to signals higher in frequency than the critical frequency?

A. They pass through the ionosphere.

B. They are absorbed by the ionosphere.

C. Their frequency is changed by the ionosphere to be below the maximum usable frequency.

D. They are reflected back to their source.

145. What is the usual input/output frequency separation for repeaters in the 2-meter band?

A. 600 kHz.

B. 1.0 MHz.

C. 1.6 MHz.

D. 5.0 MHz.

146. Which band may not be used by earth stations for satellite communications?

A. 6 m.

B. 2 m.

C. 70 cm.

D. 23 cm.

147. What is the usual bandwidth of a single-sideband amateur signal?

A. 1 kHz.

B. 2 kHz.

C. Between 3 and 6 kHz.

D. Between 2 and 3 kHz.

148. What is the maximum frequency shift permitted for RTTY or data transmission below 50 MHz?

A. 0.1 kHz.

B. 0.5 kHz.
C. 1 kHz.
D. 5 kHz.

149. What is the maximum authorized bandwidth of RTTY, data, or multiplexed emissions using an unspecified digital code within the frequency range of 50 to 222 MHz?

A. 20 kHz.
B. 50 kHz.
C. The total bandwidth shall not exceed that of a single-sideband phone emission.
D. The total bandwidth shall not exceed 10 times that of a CW emission.

150. What is the maximum symbol rate permitted for RTTY or data transmission between 50 and 222 MHz?

A. 56 kilobaud.
B. 19.6 kilobaud.
C. 1200 baud.
D. 300 baud.

151. What is the maximum authorized bandwidth of RTTY, data, or multiplexed emissions using an unspecified digital code within the 70-cm amateur band?

A. 300 kHz.
B. 200 kHz.
C. 100 kHz.
D. 50 kHz.

152. What is the maximum frequency shift permitted for RTTY or data transmissions above 50 MHz?

A. 0.1 kHz or the sending speed, in baud, whichever is greater.
B. 0.5 kHz or the sending speed, in baud, whichever is greater.
C. 5 kHz or the sending speed, in baud, whichever is greater.
D. The FCC rules do not specify a maximum frequency shift above 50 MHz.

153. What is the maximum authorized bandwidth of RTTY, data, or multiplexed emissions using an unspecified digital code within the frequency range of 222 to 450 MHz?

A. 50 kHz.
B. 100 kHz.
C. 150 kHz.
D. 200 kHz.

154. What is the maximum symbol rate permitted for RTTY or data transmissions above 222 MHz?

   A. 300 baud.
   B. 1200 baud.
   C. 19.6 kilobaud.
   D. 56 kilobaud.

155. What is the maximum symbol rate permitted for RTTY or data transmission between 50 and 222 MHz?

   A. 56 kilobaud.
   B. 19.6 kilobaud.
   C. 1200 baud.
   D. 300 baud.

156. If the FCC rules say that the amateur service is a secondary user of a frequency band, and another service is a primary user, what does this mean?

   A. Amateurs must increase transmitter power to overcome any interference caused by primary users.
   B. Amateurs are allowed to use the frequency band only if they do not cause harmful interference to primary users.
   C. Amateurs are only allowed to use the frequency band during emergencies.
   D. Nothing special; all users of a frequency band have equal rights to operate.

157. If you let an unlicensed third-party use your amateur station, what must you do at your station's control point?

   A. You must continuously monitor and supervise the third party's participation.
   B. You must monitor and supervise the communication only if contacts are made in countries which have no third-party communications agreement with the U.S.
   C. You must monitor and supervise the communication only if contacts are made on frequencies below 30 MHz.
   D. You must key the transmitter and make the station identification.

158. What is the maximum symbol rate permitted for packet transmissions on the 2-meter band?

   A. 300 baud.
   B. 1200 baud.
   C. 19.6 kilobaud.
   D. 56 kilobaud.

159. What are the station identification requirements for an amateur transmitter used for telecommand (control) of model craft?

   A. Station identification is not required if the transmitter is labeled with the station licensee's name, address and callsign.
   B. At the beginning and end of each transmission.
   C. Once every ten minutes, and at the beginning and end of each transmission.
   D. Once every ten minutes.

160. What emission type may always be used for station identification, regardless of the transmitting frequency?

    A. CW.
    B. RTTY.
    C. MCW.
    D. Phone.

161. If a repeater is causing harmful interference to another repeater and a frequency coordinator has not recommended either station, who is primarily responsible for resolving the interference?

    A. Both repeater licensees.
    B. The licensee of the repeater which has been in operation for the longest period of time.
    C. The licensee of the repeater which has been in operation for the shortest period of time.
    D. The frequency coordinator.

162. What do the FCC rules suggest you use as an aid for correct station identification when using phone?

    A. A speech compressor.
    B. Q signals.
    C. A phonetic alphabet.
    D. Unique words of your choice.

163. Which operator licenses authorize privileges on 52.525 MHz?

    A. Extra, Advanced only.
    B. Extra, Advanced, General only.
    C. Extra, Advanced, General, Technician only.
    D. Extra, Advanced, General, Technician, Novice.

164. If a disaster disrupts normal communication systems in an area where the amateur service is regulated by the FCC, what kinds of transmissions may stations make?

    A. Those that are to be used for program production or newsgathering for broadcasting purposes.
    B. Those for which material compensation has been paid to the amateur operator for delivery into the affected area.
    C. Those that allow a commercial business to continue to operate in the affected area.
    D. Those that are necessary to meet essential communication needs and facilitate relief actions.

165. What are messages called that are sent into or out of a disaster area concerning the immediate safety of human life?

A. Tactical traffic.

B. Emergency traffic.

C. Formal message traffic.

D. Health and Welfare traffic.

166. What must you do to renew or change your operator/primary station license?

A. Properly fill out FCC Form 610 and send it to the FCC in Gettysburg, PA.

B. Properly fill out FCC Form 610 and send it to the nearest FCC field office.

C. Properly fill out FCC Form 610 and send it to the FCC in Washington, DC.

D. An amateur license never needs changing or renewing.

167. What is the "grace period" during which the FCC will renew an expired 10-year license?

A. There is no grace period.

B. 10 years.

C. 5 years.

D. 2 years.

168. When are third-party messages allowed to be sent to a foreign country?

A. When sent by agreement of both control operators.

B. When the third party speaks to a relative.

C. They are not allowed under any circumstances.

D. When the U.S. has a third-party agreement with the foreign country or the third party is qualified to be a control operator.

169. If you are using a language besides English to make a contact, what language must you use when identifying your station?

A. The language being used for the contact.

B. The language being used for the contact, providing the U.S. has a third-party communications agreement with that country.

C. English.

D. Any language of a country which is a member of the International Telecommunication Union.

170. What is the proper distress call to use when operating phone?

A. Say "mayday" several times.

B. Say "help" several times.

C. Say "emergency" several times.

D. Say "SOS" several times.

171. If a repeater is causing harmful interference:

A. The licensee of the unrecommended repeater.

B. Both repeater licensees.

C. The licensee of the recommended repeater.

D. The frequency coordinator.

172. If you wanted to use your amateur station to retransmit communications between a space shuttle and its associated earth stations, what agency must first give its approval?

    A. The FCC in Washington, DC.
    B. The office of your local FCC Engineer in Charge (EIC).
    C. The National Aeronautics and Space Administration.
    D. The Department of Defense.

173. If you are a Technician licensee, what must you have to prove that you are authorized to use the Novice amateur frequencies below 30 MHz?

    A. A certificate from the FCC showing that you have notified them that you will be using the HF bands.
    B. A certificate from an instructor showing that you have attended a class in HF communications.
    C. Written proof of having passed a Morse code test.
    D. No special proof is required before using the HF bands.

174. Which list of emission types is in order from the narrowest bandwidth to the widest bandwidth?

    A. RTTY, CW, SSB voice, FM voice.
    B. CW, FM voice, RTTY, SSB voice.
    C. CW, RTTY, SSB voice, FM voice.
    D. CW, SSB voice, RTTY, FM voice.

175. What is the proper distress call to use when operating CW?

    A. Mayday.
    B. QRRR.
    C. QRZ.
    D. SOS.

176. According to the ANSI RF protection guide, what frequencies cause us the greatest risk from RF energy?

    A. Above 1500 MHz.
    B. 300 to 3000 MHz.
    C. 30 to 300 MHz.
    D. 3 to 30 MHz.

177. What does a directional wattmeter measure?

    A. Forward and reflected power.
    B. The directional pattern of an antenna.
    C. The energy used by a transmitter.
    D. Thermal heating in a load resistor.

178. Which body organ can be fatally affected by a very small amount of electrical current?

    A. The liver.

    B. The heart.

    C. The brain.

    D. The lungs.

179. How much electrical current flowing through the human body is usually painful?

    A. As little as $\frac{1}{500}$ of an ampere.

    B. Approximately 10 amperes.

    C. More than 20 amperes.

    D. Current flow through the human body is never painful.

180. What is the purpose of the ANSI RF protection guide?

    A. It lists all RF frequency allocations for interference protection.

    B. It gives RF exposure limits for the human body.

    C. It sets transmitter power limits for interference protection.

    D. It sets antenna height limits for aircraft protection.

181. Where should the main power switch for a high-voltage power supply be located?

    A. A high-voltage power supply should not be switch-operated.

    B. Anywhere that can be seen and reached easily.

    C. On the back side of the cabinet, out of sight.

    D. Inside the cabinet, to kill the power if the cabinet is opened.

182. What is the most important accessory to have for a hand-held radio in an emergency?

    A. An extra antenna.

    B. A portable amplifier.

    C. Several sets of charged batteries.

    D. A microphone headset for hands-free operation.

183. What is an important safety rule concerning the main electrical box in your home?

    A. Make sure the door cannot be opened easily.

    B. Make sure something is placed in front of the door so no one will be able to get to it easily.

    C. Make sure others in your home know where it is and how to shut off the electricity.

    D. Warn others in your home never to touch the switches, even in an emergency.

184. According to the ANSI RF protection guide, what is the maximum safe power output to the antenna of a hand-held VHF or UHF radio?

    A. 125 mW.

B. 7 W.

C. 10 W.

D. 25 W.

185. What is used to measure relative signal strength in a receiver?

A. An S meter.

B. An RST meter.

C. A signal deviation meter.

D. An SSB meter.

186. How can exposure to a large amount of RF energy affect body tissue?

A. It causes radiation poisoning.

B. It heats the tissue.

C. It paralyzes the tissue.

D. It produces genetic changes in the tissue.

187. Which body organ is the most likely to be damaged from the heating effects of RF radiation?

A. Eyes.

B. Hands.

C. Heart.

D. Liver.

188. What other emission does phase modulation most resemble?

A. Frequency modulation.

B. Pulse modulation.

C. Amplitude modulation.

D. Single-sideband modulation.

189. If you use a 3- to 30-MHz RF power meter for VHF, how accurate will its readings be?

A. They will not be accurate.

B. They will be accurate enough to get by.

C. If it properly calibrates to full scale in the set position, they may be accurate.

D. They will be accurate providing the readings are multiplied by 4.5.

190. What is a marker generator?

A. A high-stability oscillator that generates reference signals at exact frequency intervals.

B. A low-stability oscillator that "sweeps" through a range of frequencies.

C. A low-stability oscillator used to inject a signal into a circuit under test.

D. A high-stability oscillator which can produce a wide range of frequencies and amplitudes.

191. If you use a 3- to 30-MHz SWR meter for VHF, how accurate will its readings be?

   A. They will not be accurate.
   B. They will be accurate enough to get by.
   C. If it properly calibrates to full scale in the set position, they may be accurate.
   D. They will be accurate providing the readings are multiplied by 4.5.

192. If a directional RF wattmeter reads 90 W forward power and 10 W reflected power, what is the actual transmitter output power?

   A. 10 W.
   B. 80 W.
   C. 90 W.
   D. 100 W.

193. At what line impedance do most RF wattmeters usually operate?

   A. 25 Ω.
   B. 50 Ω.
   C. 100 Ω.
   D. 300 Ω.

194. How is a marker generator used?

   A. To calibrate the tuning dial on a receiver.
   B. To calibrate the volume control on a receiver.
   C. To test the amplitude linearity of a transmitter.
   D. To test the frequency deviation of a transmitter.

195. How can the range of an ammeter be increased?

   A. By adding resistance in series with the circuit under test.
   B. By adding resistance in parallel with the circuit under test.
   C. By adding resistance in series with the meter.
   D. By adding resistance in parallel with the meter.

196. How might you check the accuracy of your receiver's tuning dial?

   A. Tune to the frequency of a shortwave broadcasting station.
   B. Tune to a popular amateur net frequency.
   C. Tune to one of the frequencies of station WWV or WWVH.
   D. Tune to another amateur station and ask what frequency the operator is using.

197. What device produces a stable, low-level signal that can be set to a desired frequency?

   A. A wavemeter.
   B. A reflectometer.
   C. A signal generator.
   D. An oscilloscope.

198. How can the range of a voltmeter be increased?

    A. By adding resistance in series with the circuit under test.
    B. By adding resistance in parallel with the circuit under test.
    C. By adding resistance in series with the meter, between the meter and the cir-cuit under test.
    D. By adding resistance in parallel with the meter, between the meter and the circuit under test.

199. What is marker generator?

    A. A high-stability oscillator that generates reference signals at exact fre-quency intervals.
    B. A low-stability oscillator that "sweeps" through a range of frequencies.
    C. A low-stability oscillator used to inject a signal into a circuit under test.
    D. A high-stability oscillator which can produce a wide range of frequencies and amplitudes.

200. What is an RF signal generator used for?

    A. Measuring RF signal amplitudes.
    B. Aligning tuned circuits.
    C. Adjusting transmitter impedance-matching networks.
    D. Measuring transmission line impedances.

201. What frequency standard can be used to calibrate the tuning dial of a receiver?

    A. A calibrated voltmeter.
    B. Signals from WWV and WWVH.
    C. A deviation meter.
    D. A sweep generator.

202. What device is used in place of an antenna during transmitter tests so that no signal is radiated?

    A. An antenna matcher.
    B. A dummy antenna.
    C. A low-pass filter.
    D. A decoupling resistor.

203. Why do many radio receivers have several IF filters of different bandwidths that can be selected by the operator?

    A. Because some frequency bands are wider than others.
    B. Because different bandwidths help increase the receiver sensitivity.
    C. Because different bandwidths improve S-meter readings.
    D. Because some emission types need a wider bandwidth than others to be re-ceived properly.

204. What circuit combines signals from an IF amplifier stage and a beat-frequency oscillator (BFO), to produce an audio signal?

A. A VFO circuit.
B. An AGC circuit.
C. A detector circuit.
D. A power supply circuit.

205. What circuit uses a limiter and a frequency discriminator to produce an audio signal?

A. A double-conversion receiver.
B. A variable-frequency oscillator.
C. A superheterodyne receiver.
D. An FM receiver.

206. What is the unlabeled block in Fig. 3-3?

A. An AGC circuit.
B. A detector.
C. A power supply.
D. A VFO circuit.

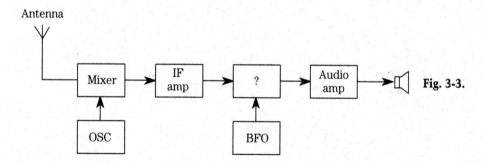

Fig. 3-3.

207. What circuit is pictured in Fig. 3-4?

A. A double-conversion receiver.
B. A variable-frequency oscillator.
C. A superheterodyne receiver.
D. An FM receiver.

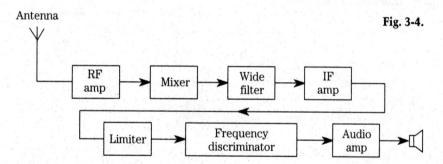

Fig. 3-4.

208. What is the meaning of: "Your signal is full quieting?"

    A. Your signal is strong enough to overcome all receiver noise.
    B. Your signal has no spurious sounds.
    C. Your signal is not strong enough to be received.
    D. Your signal is being received, but no audio is being heard.

209. Which ionospheric region limits daytime radio communications on the 80-meter band to short distances?

    A. D region.
    B. E region.
    C. F1 region.
    D. F2 region.

210. Which ionospheric region is closest to the earth?

    A. The A region.
    B. The D region.
    C. The E region.
    D. The F region.

211. When is the ionosphere most ionized?

    A. Dusk.
    B. Midnight.
    C. Midday.
    D. Dawn.

212. What effect does the D region of the ionosphere have on lower-frequency HF signals in the daytime?

    A. It absorbs the signals.
    B. It bends the radio waves out into space.
    C. It refracts the radio waves back to earth.
    D. It has little or no effect on 80-meter radio waves.

213. What is the result of overdeviation in an FM transmitter?

    A. Increased transmitter power.
    B. Out-of-channel emissions.
    C. Increased transmitter range.
    D. Poor carrier suppression.

214. What is the unlabeled block in Fig. 3-5?

    A. A bandpass filter.
    B. A crystal oscillator.
    C. A reactance modulator.
    D. A rectifier modulator.

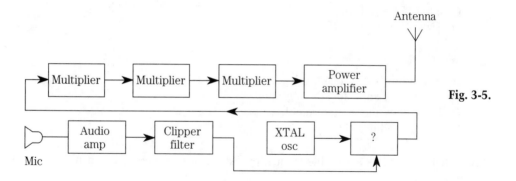

Fig. 3-5.

215. What is modulation?

    A. Varying a radio wave in some way to send information.
    B. Receiving audio information from a signal.
    C. Increasing the power of a transmitter.
    D. Suppressing the carrier in a single-sideband transmitter.

216. What emissions are produced by a transmitter using a reactance modulator?

    A. CW.
    B. Test.
    C. Single-sideband, suppressed-carrier phone.
    D. Phase-modulated phone.

217. How would you modulate a 2-meter FM transceiver to produce packet-radio emissions?

    A. Connect a terminal-node-controller to interrupt the transceiver's carrier wave.
    B. Connect a terminal-node-controller to the transceiver's microphone input.
    C. Connect a keyboard to the transceiver's microphone input.
    D. Connect a DTMF keypad to the transceiver's microphone input.

218. What is the name for Morse code emissions produced by switching a transmitter's output on and off?

    A. Phone.
    B. Test.
    C. CW.
    D. RTTY.

219. What is the maximum transmitting power permitted an amateur station in beacon operation?

    A. 10 W PEP output.
    B. 100 W PEP output.

C. 500 W PEP output.

D. 1500 W PEP output.

220. What is RTTY?

 A. Amplitude-keyed telegraphy.

 B. Frequency-shift-keyed telegraphy.

 C. Frequency-modulated telephony.

 D. Phase-modulated telephony.

221. What circuit has a variable-frequency oscillator connected to a driver and a power amplifier?

 A. A packet-radio transmitter.

 B. A crystal-controlled transmitter.

 C. A single-sideband transmitter.

 D. A VFO-controlled transmitter.

222. How can on-the-air interference be minimized during a lengthy transmitter testing or loading up procedure?

 A. Choose an unoccupied frequency.

 B. Use a dummy load.

 C. Use a non-resonant antenna.

 D. Use a resonant antenna that requires no loading-up procedure.

223. What is meant by the term *broadcasting*?

 A. Transmissions intended for reception by the general public, either direct or relayed.

 B. Retransmission by automatic means of programs or signals from nonamateur stations.

 C. One-way radio communications, regardless of purpose or content.

 D. One-way or two-way radio communications between two or more stations.

224. How is tone-modulated Morse code produced?

 A. By on/off keying of a carrier.

 B. By feeding an on/off keyed audio tone into a CW transmitter.

 C. By feeding an on/off keyed audio tone into a transmitter.

 D. By feeding a microphone's audio signal into an FM transmitter.

225. What is the name for packet-radio emissions?

 A. CW.

 B. Data.

 C. Phone.

 D. RTTY.

226. Why do modern HF transmitters have a built-in low-pass filter in their RF output circuits?

    A. To reduce RF energy below a cutoff point.

    B. To reduce low-frequency interference to other amateurs.

    C. To reduce harmonic radiation.

    D. To reduce fundamental radiation.

227. What is the control point of an amateur station?

    A. The on/off switch of the transmitter.

    B. The input/output port of a packet controller.

    C. The variable frequency oscillator of a transmitter.

    D. The location at which the control operator function is performed.

228. Why is it a good idea to have a way to operate your amateur station without using commercial ac power lines?

    A. So you may use your station while mobile.

    B. So you may provide communications in an emergency.

    C. So you may operate in contests where ac power is not allowed.

    D. So you will comply with the FCC rules.

229. What circuit is pictured in Fig. 11-2 if block 1 is a variable-frequency oscillator?

    A. A packet-radio transmitter.

    B. A crystal-controlled transmitter.

    C. A single-sideband transmitter.

    D. A VFO-controlled transmitter.

230. What kind of emission would your FM transmitter produce if its microphone failed to work?

    A. An unmodulated carrier.

    B. A phase-modulated carrier.

    C. An amplitude-modulated carrier.

    D. A frequency-modulated carrier.

231. What is a courtesy tone (used in repeater operations)?

    A. A sound used to identify the repeater.

    B. A sound used to indicate when a transmission is complete.

    C. A sound used to indicate that a message is waiting for someone.

    D. A sound used to activate a receiver in case of severe weather.

232. What type of amateur station transmits communications for the purpose of observation of propagation and reception?

    A. A beacon.

    B. A repeater.
    C. An auxiliary station.
    D. A radio control station.

233. What causes a repeater to "time out?"

    A. The repeater's battery supply runs out.
    B. Someone's transmission goes on longer than the repeater allows.
    C. The repeater gets too hot and stops transmitting until its circuitry cools off.
    D. Something is wrong with the repeater.

234. Ducting occurs in which region of the atmosphere?

    A. F2.
    B. Ectosphere.
    C. Troposphere.
    D. Stratosphere.

235. Why is FM voice best for local VHF/UHF radio communications?

    A. The carrier is not detectable.
    B. It is more resistant to distortion caused by reflected signals.
    C. It has high-fidelity audio which can be understood even when the signal is somewhat weak.
    D. Its RF carrier stays on frequency better than the AM modes.

236. What is the name for emissions produced by an on/off keyed audio tone?

    A. RTTY.
    B. MCW.
    C. CW.
    D. Phone.

237. What is the purpose of repeater operation?

    A. To cut your power bill by using someone else's higher power system.
    B. To help mobile and low-power stations extend their usable range.
    C. To transmit signals for observing propagation and reception.
    D. To make calls to stores more than 50 miles away.

238. What is the name of the area that makes long-distance radio communications possible by bending radio waves?

    A. Troposphere.
    B. Stratosphere.
    C. Magnetosphere.
    D. Ionosphere.

239. How are VHF signals propagated within the range of the visible horizon?

    A. By sky wave.

B. By direct wave.

C. By plane wave.

D. By geometric wave.

240. What causes the ionosphere to absorb radio waves?

    A. The weather below the ionosphere.

    B. The ionization of the D region.

    C. The presence of ionized clouds in the E region.

    D. The splitting of the F region.

241. What effect does tropospheric bending have on 2-meter radio waves?

    A. It lets you contact stations farther away.

    B. It causes them to travel shorter distances.

    C. It garbles the signal.

    D. It reverses the sideband of the signal.

242. When does ionospheric absorption of radio signals occur?

    A. When tropospheric ducting occurs.

    B. When long wavelength signals enter the D region.

    C. When signals travel to the F region.

    D. When a temperature inversion occurs.

243. What band conditions might indicate long-range skip on the 6-meter and 2-meter bands?

    A. Noise on the 80-meter band.

    B. The absence of signals on the 10-meter band.

    C. Very long-range skip on the 10-meter band.

    D. Strong signals on the 10-meter band from stations about 500 to 600 miles away.

244. Which region of the ionosphere is mainly responsible for absorbing radio signals during the daytime?

    A. The D region.

    B. The E region.

    C. The F1 region.

    D. The F2 region.

245. Which region of the ionosphere is mainly responsible for long-distance sky-wave radio communications?

    A. D region.

    B. E region.

    C. F1 region.

    D. F2 region.

246. Which region of the ionosphere is the least useful for long-distance radio wave propagation?

    A. The D region.
    B. The E region.
    C. The F1 region.
    D. The F2 region.

247. When is the E region most ionized?

    A. Dawn.
    B. Midday.
    C. Dusk.
    D. Midnight.

248. Which ionospheric region most affects sky-wave propagation on the 6-meter band?

    A. The D region.
    B. The E region.
    C. The F1 region.
    D. The F2 region.

249. What causes the ionosphere to form?

    A. Solar radiation ionizing the outer atmosphere.
    B. Temperature changes ionizing the outer atmosphere.
    C. Lightning ionizing the outer atmosphere.
    D. Release of fluorocarbons into the atmosphere.

250. Which ionospheric region is closest to earth.

    A. The A region.
    B. The D region.
    C. The E region.
    D. The F region.

251. What causes VHF radio waves to be propagated several hundred miles over oceans?

    A. A widespread temperature inversion.
    B. An overcast of cirriform clouds.
    C. A high-pressure zone.
    D. A polar air mass.

252. What kind of propagation would best be used by two stations within each other's skip zone on a certain frequency?

    A. Ground-wave.
    B. Sky-wave.
    C. Scatter-mode.
    D. Ducting.

253. When is the ionosphere least ionized?

    A. Shortly before dawn.
    B. Just after noon.
    C. Just after dusk.
    D. Shortly before midnight.

254. What two sub-regions of ionosphere exist only in the daytime?

    A. Troposphere and stratosphere.
    B. F1 and F2.
    C. Electrostatic and electromagnetic.
    D. D and E.

255. What effect does the D region of the ionosphere have on lower-frequency HF signals in the daytime?

    A. It has little or no effect on 80-meter radio waves.
    B. It refracts the radio waves back to earth.
    C. It bends the radio waves out into space.
    D. It absorbs the signals.

256. What weather condition can cause tropospheric ducting?

    A. Periods of heavy rainfall.
    B. A series of low-pressure waves.
    C. An unstable low-pressure system.
    D. A stable high-pressure system.

257. Which region of the ionosphere is mainly responsible for long-distance sky-wave radio communications?

    A. D region.
    B. E region.
    C. F1 region.
    D. F2 region.

258. If you are receiving a weak and distorted signal from a distant station on a frequency that is close to the maximum usable frequency, what type of propagation is probably occurring?

    A. Ducting.
    B. Line-of-sight.
    C. Scatter.
    D. Ground-wave.

259. Which two daytime ionospheric regions combine into one region at night?

    A. E and F1.
    B. D and E.
    C. F1 and F2.
    D. E1 and E2.

260. What is the name for unmodulated carrier wave emissions?

    A. Phone.
    B. Test.
    C. CW.
    D. RTTY.

261. What does vertical wave polarization mean?

    A. The electric lines of force of a radio wave are parallel to the earth's surface.
    B. The magnetic lines of force of a radio wave are perpendicular to the earth's surface.
    C. The electric lines of force of a radio wave are perpendicular to the earth's surface.
    D. The electric and magnetic lines of force of a radio wave are parallel to the earth's surface.

262. What is the name of the voice emission most used on VHF/UHF repeaters?

    A. Single-sideband phone.
    B. Pulse-modulated phone.
    C. Slow-scan phone.
    D. Frequency modulated phone.

263. Why should local amateur communications use VHF and UHF frequencies instead of HF frequencies?

    A. To minimize interference on HF bands capable of long distance communication.
    B. Because greater output power is permitted on VHF and UHF.
    C. Because HF transmissions are not propagated locally.
    D. Because signals are louder on VHF and UHF frequencies.

264. What is the maximum symbol rate permitted for RTTY or data transmissions between 38 and 50 MHz?

    A. 56 kilobauds.
    B. 19.6 kilobauds.
    C. 1200 bauds.
    D. 300 bauds.

265. What type of message concerning a person's well-being are sent into or out of a disaster area?

    A. Health and Welfare traffic.
    B. Formal message traffic.
    C. Tactical traffic.
    D. Routine Traffic.

266. Which list of emission types is in order from the narrowest bandwidth to the widest bandwidth?

A. CW, SSB voice, RTTY, FM voice.
B. CW, FM voice, RTTY, SSB voice.
C. RTTY, CW, SSB voice, FM voice.
D. AM, CW, RTTY.

267. What is the name of the voice emission most used on amateur HF bands?

A. Single-sideband phone.
B. Pulse-modulated phone.
C. Slow-scan phone.
D. Frequency modulated phone.

268. If you are talking to a station using a repeater, how would you find out if you could communicate using simplex instead?

A. See if you can clearly receive the station on the repeater's input frequency.
B. See if you can clearly receive the station on a lower frequency band.
C. See if you can clearly receive a more distant receiver.
D. See if a third station can clearly receive both of you.

269. What is the proper *Q* signal to use to ask if someone is calling you on CW?

A. QRT.
B. QRL.
C. QSL.
D. QRZ.

270. What is the meaning of: "Your signal report is three by three . . ."?

A. The contact is serial number 33.
B. The station is located at latitude 33 degrees.
C. Your signal is readable with considerable difficulty and weak in strength.
D. Your signal is unreadable and very weak in strength.

271. During commuting rush hours, which type of repeater operation should be discouraged?

A. Mobile stations.
B. Low-power stations.
C. Highway traffic information nets.
D. Third-party communications nets.

272. What is the meaning of: "Your signal report is five by nine plus 20 dB?"

A. A relative signal-strength meter reading is 20 decibels greater than strength 9.
B. The bandwidth of your signal is 20 decibels above linearity.
C. Repeat your transmission on a frequency 20 kHz higher.
D. Your signal strength has increased by a factor of 100.

273. What is the term for the location at which the control operator function is performed?

    A. The control point.
    B. The station point.
    C. The manual control location.
    D. The operating desk.

274. What organization has published safety guidelines for the maximum limits of RF energy near the human body?

    A. The American National Standards Institute (ANSI).
    B. The Environmental Protection Agency (EPA).
    C. The Federal Communications Commission (FCC).
    D. The Institute of Electrical and Electronics Engineers (IEEE).

275. What is one meaning of the Q signal "QSY"?

    A. Change frequency.
    B. Send more slow.
    C. Send faster.
    D. Use more power.

276. What causes splatter interference?

    A. Overmodulation of a transmitter.
    B. Keying a transmitter too fast.
    C. Signals from a transmitter's output circuit are being sent back into its input circuit.
    D. The transmitting antenna is the wrong length.

277. How do you call another station on a repeater if you know the station's callsign?

    A. Say "break, break 79," then say the station's callsign.
    B. Say the station's callsign, then identify your own station.
    C. Say "CQ" three times, then say the station's callsign.
    D. Wait for the station to call "CQ," then answer it.

278. What is a repeater frequency coordinator?

    A. Someone who organizes the assembly of a repeater station.
    B. Someone who provides advice on what kind of repeater to buy.
    C. The person whose callsign is used for a repeater's identification.
    D. A person or group that recommends frequencies for repeater usage.

279. What is the proper way to interrupt a repeater conversation to signal a distress call?

    A. Say "BREAK" twice, then your callsign.
    B. Say "HELP" as many times as it takes to get someone to answer.
    C. Say "SOS," then your callsign.
    D. Say "EMERGENCY" three times.

280. How might you join a closed repeater system?

    A. Contact the control operator and ask to join.
    B. Use the repeater until told not to.
    C. Use simplex on the repeater input until told not to.
    D. Write the FCC and report the closed condition.

281. What is the proper way to break into a conversation on a repeater?

    A. Say your callsign during a break between transmissions.
    B. Turn on an amplifier and override whoever is talking.
    C. Shout, "break, break!" to show that you're eager to join the conversation.
    D. Wait for the end of a transmission and start calling the desired party.

282. What is the proper Q signal to use to see if a frequency is in use before transmitting on CW?

    A. QRV.
    B. QRU.
    C. QRL.
    D. QRZ.

283. What is a repeater called that is available for anyone to use?

    A. An open repeater.
    B. A closed repeater.
    C. An autopatch repeater.
    D. A private repeater.

284. Why should simplex be used where possible, instead of using a repeater?

    A. Signal range will be increased.
    B. Long distance toll charges will be avoided.
    C. The repeater will not be tied up unnecessarily.
    D. Your antenna's effectiveness will be better tested.

285. If you are operating simplex on a repeater frequency, why would it be good amateur practice to change to another frequency?

    A. The repeater's output power may ruin your station's receiver.
    B. There are more repeater operators than simplex operators.
    C. Changing the repeater's frequency is not practical.
    D. Changing the repeater's frequency requires the authorization of the FCC.

286. How should you give a signal report over a repeater?

    A. Say what your receiver's S-meter reads.
    B. Always say: "Your signal report is five five . . ."
    C. Say the amount of signal quieting into the repeater.
    D. Try to imitate the sound quality you are receiving.

287. What is the proper way to ask someone their location when using a repeater?
    A. Where are you?
    B. What is your QTH?
    C. What is your 20?
    D. Locations are not normally told by radio.

288. What is one reason for using tactical callsigns, such as "command post" or "weather center" during an emergency?
    A. They keep the general public informed about what is going on.
    B. They are more efficient and help coordinate public-service communications.
    C. They are required by the FCC.
    D. They increase goodwill between amateurs.

289. Why should you pause briefly between transmissions when using a repeater?
    A. To check the SWR of the repeater.
    B. To reach for pencil and paper for third-party communications.
    C. To listen for anyone wanting to break in.
    D. To dial up the repeater's autopatch.

290. What is one meaning of the Q signal "QSO"?
    A. A contact is confirmed.
    B. A conversation is in progress.
    C. A contact is ending.
    D. A conversation is desired.

291. Why should you keep transmissions short when using a repeater?
    A. A long transmission might prevent someone with an emergency from using the repeater.
    B. To see if the receiving station operator is still awake.
    C. To give any listening nonhams a chance to respond.
    D. To keep long distance charges down.

Chapter 3
Answer sheet

Questions and answers were supplied by the FCC.

1. B	2. A	3. C	4. C	5. A
6. A	7. D	8. A	9. A	10. B
11. A	12. D	13. D	14. B	15. A
16. D	17. B	18. B	19. C	20. C
21. A	22. D	23. C	24. A	25. C
26. A	27. B	28. B	29. A	30. D

31. A	32. C	33. C	34. B	35. A
36. A	37. B	38. A	39. D	40. A
41. B	42. C	43. D	44. D	45. A
46. D	47. B	48. B	49. D	50. D
51. A	52. D	53. B	54. B	55. C
56. A	57. D	58. A	59. A	60. A
61. A	62. A	63. C	64. B	65. B
66. D	67. C	68. A	69. A	70. B
71. A	72. C	73. B	74. D	75. B
76. C	77. D	78. A	79. D	80. B
81. A	82. C	83. D	84. C	85. A
86. A	87. C	88. D	89. C	90. B
91. A	92. C	93. C	94. B	95. C
96. A	97. D	98. D	99. D	100. B
101. D	102. D	103. A	104. C	105. A
106. D	107. D	108. A	109. B	110. A
111. B	112. A	113. B	114. B	115. C
116. A	117. D	118. B	119. D	120. B
121. D	122. D	123. A	124. C	125. B
126. D	127. C	128. D	129. B	130. D
131. A	132. C	133. C	134. A	135. C
136. C	137. B	138. A	139. A	140. A
141. D	142. A	143. C	144. A	145. A
146. A	147. D	148. C	149. A	150. B
151. C	152. D	153. B	154. D	155. B
156. B	157. A	158. C	159. A	160. A
161. A	162. C	163. C	164. D	165. B
166. A	167. D	168. D	169. C	170. A
171. A	172. C	173. C	174. C	175. D
176. C	177. A	178. B	179. A	180. B
181. B	182. C	183. C	184. B	185. A
186. B	187. A	188. A	189. A	190. A
191. C	192. B	193. B	194. A	195. D
196. C	197. C	198. C	199. A	200. B
201. B	202. B	203. D	204. C	205. D
206. B	207. D	208. A	209. A	210. B
211. C	212. A	213. B	214. C	215. A
216. D	217. B	218. C	219. B	220. B
221. D	222. B	223. A	224. C	225. B
226. C	227. D	228. B	229. D	230. A
231. B	232. A	233. B	234. C	235. C
236. B	237. B	238. D	239. B	240. B
241. A	242. B	243. D	244. A	245. D
246. A	247. B	248. B	249. A	250. B
251. A	252. C	253. A	254. B	255. D
256. D	257. D	258. C	259. C	260. B

261. C	262. D	263. A	264. C	265. A
266. A	267. A	268. A	269. D	270. C
271. D	272. A	273. A	274. A	275. A
276. A	277. B	278. D	279. D	280. A
281. A	282. C	283. A	284. C	285. C
286. C	287. A	288. B	289. C	290. B
291. A				

4
CHAPTER

Reference material for Marine (Maritime) Radio Operator Permit

This chapter, and the information in chapter 5, (next chapter) contains all of the information you need to know in order to pass the FCC examination for the Marine Radio Operator Permit.

Included in this chapter is a list of topics covered by the examination and a list of questions similar to the questions on the examination. At the end of each topic and question is a reference to the place in reference material where the subject is discussed.

Stations requiring Marine Radio Operator Permits

A Marine Radio Operator Permit (or higher class of license) is required for persons who operate a radiotelephone (voice) station:

- on a cargo ship of 300 or more gross tons that is navigated on the open sea; or

- on a vessel sailing the Great Lakes that is more than 65 feet long or towing another vessel that is more than 65 feet long, or is carrying more than six passengers for hire; or

- at a coast station not located in Alaska that uses frequencies lower than 30 MHz and that uses 250 watts or less of carrier power or 1500 watts or less of peak envelope power; or

- on a vessel carrying more than six passengers for hire (regardless of six) that is navigated in the open sea or in any tidewaters adjacent to the open sea more than 1000 feet from shore; or

- at a coast station in Alaska which uses frequencies lower than 30 MHz and uses more than 250 watts of carrier power or more than 1500 watts of peak envelope power.

The Marine Radio Operator Permit can also be used to operate any class of station that can be operated by the holder of a Restricted Radiotelephone Operator Permit (except broadcast stations).

The Marine Radio Operator Permit cannot be used at coast or ship stations, located other than in Alaska, which transmit with more than 250 watts of carrier power or more than 1500 watts of peak envelope power.

Rules and regulations

The following paragraphs summarize various laws and regulations governing the use of radio with which you, as a licensed radio operator, should be familiar. References to specific rule sections or laws are given in parentheses. This chapter is not a substitute for the laws and regulations covered; it is intended only to consolidate in concise, understandable form, the information needed to pass the FCC Marine Radio Operator Permit examination. The abbreviations used in this chapter are as follows:

- *FCC Rule*   FCC Rules and Regulations
- *47 U. S. C.*   The Communications Act of 1934, as amended
- *IRR*   International Radio Regulations.

It is not legal to operate a radio station within the United States, including its territories, possessions, and the District of Columbia, unless that radio station is operated in accordance within the Communications Act of 1934, and the terms of a license granted under the provisions of that Act. (Section 301) This means that you must not operate a radio station unless that station is properly licensed by the FCC. Willful or repeated operation of an unlicensed radio station in violation of FCC rules and/or the Act can be punished by fines or imprisonment.

Except for broadcasts intended for public reception, amateur or citizens band communications, or distress messages, it is not legal to disclose the existence or content of any radio communication to anyone, except the party to whom the communication is addressed. [47 U. S. C. Section 705 (a)] This means that if you are sending private messages for someone on your ship (or supervising a radiotelephone conversation by a passenger or crew member), you must not reveal to anyone other than the coast or ship station communicated with, the existence or content of that communication. If you happen to overhear a radio communication that is not addressed to you, you must not tell anyone what you heard unless the communication was from an aircraft or a ship in distress. Furthermore, it is illegal for you to use the content of any intercepted communication that you receive for your own benefit, or for the benefit of someone else who is not entitled to receive it. (FCC Rule 80.88)

It is illegal for a licensed radio operator to violate or cause, aid or abet the violation or any Act, treaty, or convention binding the U. S. that the Commission is authorized to administer, or any regulation made by the Commission under such Act, treaty, or convention. (FCC Rule 13.62)

Marine Radio Operator Permits are issued by the FCC, normally for a term of five years from the date of issuance. (FCC Rule 13.4) Detailed information about application forms and filing procedures for commercial radio licenses is contained in the bulletin "Commercial Radio Operator Licenses and Permits," which is available from any commission office. (FCC Rule 13.11)

It is necessary to apply for renewal of a Marine Radio Operator Permit during the last year of its five-year term, or during a five-year period of grace after the permit expires. Expired permits are not valid. (FCC Rule 13.28)

Normally, you must post your Marine Radio Operator Permit at the place where you are on duty as a radio operator. (FCC Rule 13.74) When applying for renewal of your Marine Radio Operator Permit, you should send that permit along with your application for renewal. (FCC Rule 13.11) While the application is pending, you should post a signed copy of the application where you normally post your Marine Radio Operator Permit. (FCC Rule 13.72)

To be eligible for a Marine Radio Operator Permit, you must be either a U. S. citizen, or an alien who is eligible for employment in the United States. (FCC Rule 13.5) You are not eligible for a Marine Radio Operator Permit if you are completely deaf or completely mute, or if for any other reason, you cannot transmit and receive spoken messages in English. (FCC Rule 13.5) There is no minimum age requirement for a Marine Radio Operator Permit.

As a licensed radio operator, you must carry out the lawful orders of the master or person lawfully in charge of the ship or aircraft on which you are employed. (47 U. S. C. Sec. 303(m)(1)(B))

It is illegal for a licensed radio operator to willfully damage, cause, or permit to be damaged, any radio apparatus or installation in any licensed radio station. (47 U. S. C. Sec. 303(m)(1)(C))

It is illegal for a licensed radio operator to transmit unnecessary, unidentified, or superfluous radio communications or signals. (47 U. S. C. Sec. 303(m)(1)(D))

Particularly in safety services, such as the maritime radio service, you must not tie up vital communications channels with idle "chit-chat." (FCC Rule 80.90)

It is illegal for a licensed radio operator to transmit communications containing obscene, indecent, or profane words, language, or meaning. (47 U. S. C. Sec. 303(m)(1)(D))

It is illegal for a licensed radio operator to transmit false or deceptive signals or communications by radio. Furthermore, it is illegal for a licensed radio operator to falsely identify a radio station by transmitting a callsign which has not been assigned by proper authority to that station. (47 U. S. C. Sec. 303(m)(1)(D))

It is illegal for a licensed radio operator to willfully or maliciously interfere with or cause interference to any radio communication or signal. (47 U. S. C. Sec. 303(m)(1)(E))

It is illegal for a licensed radio operator to alter, duplicate for fraudulent purposes, fraudulently obtain, or assist another to alter, duplicate for fraudulent purposes, or fraudulently obtain an operator license. (FCC Rule 13.70)

If your Marine Operator Permit gets lost, mutilated, or destroyed, you should apply for a duplicate permit, including an explanation of what happened to it. In the event that you later find the original permit, you must return it, (or the duplicate permit) to the FCC for cancellation. (FCC Rule 13.71)

Any licensee who appears to have violated any provision of the Communications Act will, before revocation, suspension, or cease and desist proceedings are instituted, be served with a written notice calling the facts to his or her attention and requesting a statement concerning the matter within ten days of receipt of the notice. FCC Form 793 is used for this purpose. (FCC Rule 1.89)

When grounds exist for suspension of an operator license (47 U. S. C. Section 303), no order of suspension shall take effect until 15 days notice in writing of the cause of the proposed suspension has been given to the operator licensee, who may make written application to the Commission at any time within said 15 days, for a hearing upon such order. Upon conclusion of said hearing, the commission may affirm, modify, or revoke said order of suspension. If the license is ordered suspended, the operator shall send his or her operator license to the office of the Commission in Washington DC, on or before the effective date of the order, or if the effective date has passed at the time the notice is received, the license shall be sent to the Commission as soon as possible thereafter. (FCC Rule 1.85)

Any person who is determined by the Commission to have:

- willfully or repeatedly failed to comply with any license, permit, certificate, or other instrument or authorization issued by the Commission, or
- failed to comply with any of the provisions of the Communications Act, or of FCC Rules and Regulations, shall be liable to the United States for a forfeiture penalty, not to exceed $2,000, for each violation. [(47 U. S. C. Section 503(b).]
- willfully violated Section 705(a) of the Communications Act shall be fined not more than $1,000 or imprisoned for not more than 6 months, or both. (47 U. S. C. Section 705(d)(1))
- willfully violated Section 705(a) of the Communications Act for purposes or direct or indirect commercial advantage or private financial gain shall be fined not more than $25,000 or imprisoned for not more than one year, or both. (47 U. S. C. Section 705(d)(2))

The Marine Radio Operator Permit is classified as a "Restricted Radiotelephone Certificate" in the International Radio Regulations. It authorizes the holder to operate certain stations in the maritime and aviation services. It does not authorize the operation of AM, FM, or TV broadcast stations. However, for the purpose of operating stations in the maritime services, the Marine Radio Operator Permit conveys all of the authority of the Restricted Radiotelephone Operator Permit. (FCC Rule 80.151)

The holder of a Marine Radio Operator Permit is authorized to operate any maritime coast radiotelephone station which transmits with 250 watts or less of carrier power, or 1500 watts or less of peak envelope power. (FCC Rule 80.153) In Alaska, the holder of a Marine Radio Operator Permit is authorized to operate any maritime coast radiotelephone station, regardless of transmitter power. (FCC Rule 80.153) A Marine Radio Operator Permit is not allowed to transmit manual telegraphy (Morse code) at a coast station, under any circumstances. (FCC Rule 80.153) Furthermore, the holder of a Marine Radio Operator Permit is not allowed to make any equipment adjustments that could result in improper transmitter operation. (FCC Rule 80.167) Marine Radio Operator Permit holders may only use transmitters that do not require manual adjustment of the frequency determining elements for routine operation. (FCC Rule 80.167)

All adjustments of radiotelephone transmitting equipment in any maritime coast or ship station, made during the installation, servicing, or maintenance of that equipment, and which may affect its proper operation, may be made only by, or under the immediate supervision of a person who holds a General Radiotelephone Operator License, or a First or Second Class Radiotelegraph Operator's Certificate. (FCC Rule 80.169)

Public coast stations using telephony are authorized to communicate:

- with any ship or aircraft station operating in the maritime mobile service for the transmission or reception of safety communication, and

- with any land station for the purpose of facilitating the transmission or reception of safety communication to or from a ship or aircraft station. (FCC Rule 80.453)

The radio operator on board a ship equipped with a radio station must allow officials of foreign governments of countries where the ship calls to examine the radio station license if they so request. Furthermore, he or she must facilitate this examination. If the license is not available, or if other irregularities are found, officials of these governments may inspect the radio station in order to satisfy themselves that complies with the International Radio Regulations. (FCC Rule 80.79) and (International Radio Regulations #4012)

The original authorization of each radio operator must be posted in a conspicuous place at the coast station or on board the ship at which the station is operated, while he or she is employed as radio operator of the station. (FCC Rule 80.175, 80.407(b))

An operator who holds a photocopy of the license, may, in lieu of posting, have the photocopy in his personal possession immediately available for inspection upon request by a Commission representative when operating the following:

- A voluntarily equipped vessel. This is a vessel which is not required by FCC Rules to have a station installed.

- Any class of ship station when the operator is on board solely for the purpose of servicing the radio equipment, or

- a station of a portable nature. (FCC Rule 80.175, 80.407(b))

It is illegal to operate a ship station's transmitter anywhere, except on that ship. Furthermore, it is illegal to use a ship station to communicate while the ship is being transported, stored, or parked on land. (FCC Rule 80.89)

You must stop transmitting immediately if you notice, or if the FCC brings to your attention, any deviation from the technical requirements for the ship or coast station. You must not transmit again until the malfunction is corrected, except for an emergency involving the immediate safety of life and property. (FCC Rule 80.90)

There are two frequencies which are internationally designated for use by maritime radiotelephone stations in distress. The medium frequency (MF) distress channel is 2182 kHz, and the very high frequency (VHF) distress channel is 156.8 MHz (also known as VHF marine Channel 16). (FCC Rule 80.369) These are the same two frequencies which are used for calling. (FCC Rule 80.369) It is illegal to use selective calling on 2182 kHz or 156.8 MHz (Channel 16). Except when making a distress call, it is illegal to transmit a general call on these frequencies; that is, a communication not addressed to a particular station or group of stations. (FCC Rule 80.89)

In the aviation radio services, it is not mandatory to hold a radio operator license unless operating on a frequency other than VHF, or on a flight to a foreign country. For practical purposes, most aviation station operators need only a Restricted Radiotelephone Operator Permit. However, a Marine Radio Operator Permit is required to operate an aircraft station using frequencies below 30 MHz not exclusively allocated to the aeronautical mobile service and which are assigned for international use. This applies to stations whose power doesn't exceed 250 watts carrier power, or 1000 watts Peak Envelope Power. If the station's power does exceed this, a General Radiotelephone Operator License is required. (FCC Rule 87.133)

Adjustments or tests relating to the installation, servicing, or maintenance of an aviation radio station, which could affect the operation of the transmitter, must be made under the direct supervision of a person who holds a General Radiotelephone Operator License, or a First or Second Class Radiotelegraph Operator's Certificate. (FCC Rules 87.135 and 80.167)

A person who holds a Marine Radio Operator Permit, a Restricted Radiotelephone Operator Permit, or a Third Class Radiotelegraph Operator's Certificate is allowed to operate only those transmitters that (1) have simple, external switching devices, and (2) do not require manual adjustment of frequency-determining elements. (FCC Rule 87.136 and 80.167)

The universal simplex clear channel frequency for use by aircraft in distress is 121.5 MHz. This frequency, and 243 MHz, is also used by shipboard EPIRB (Emergency Position Indicating Radio Beacon) station. (FCC Rule 80.1053)

No radio operator license is required for the operation of transmitters involving:

- The operation of airborne radar sets, radio altimeters, transponders, and other airborne automatic radio navigation aids,
- Operation of Emergency Locator Transmitters (ELT's),
- Operation of EPIRB transmitters,
- Operation of any authorized radio station which retransmits communications by automatic means,
- Operation of any aeronautical enroute station which transmits, by automatic means, digital communications to aircraft station, or
- Operation of VHF telephony transceivers which provide domestic service or, are used on domestic flights. (FCC Rule 87.139)

Caution: Although no operator authorization is required for the above, the transmitters must be covered by a valid station license or other authorization.

Operating procedures

The most important practice that a radio operator must learn is to monitor the channel before transmitting. By doing this, he or she reduces the possibility of interfering with other stations which may be using the channel.

Two-way communications systems are operated in either a simplex or a duplex mode. *Duplex* means that both persons can talk at the same time, just like on the phone. *Simplex* means that only one of two persons can talk at a time, so that each must wait his or her turn to speak. Most public coast stations which connect with land-line telephone operate in the duplex mode. When using your marine radio in a simplex mode, you must wait until all other stations stop transmitting before you transmit. Otherwise, you will interfere with the other stations using the channel and neither you, nor the stations you interfere with will be heard.

When communicating using a simplex radio system, you should pause for a few seconds before replying to another station's transmissions. This interval will allow another station to "break-in," if necessary. When communications are in progress on a channel, a radio operator should break in only when there is an emergency, or to call another station briefly and then move to a channel that is not busy.

Normally, a radio operator using simplex will say "over" when he or she intends to turn over the channel to the other station, and is expecting a reply. If a radio operator is finished using the channel and does not expect a reply, he or she will say "out" or "clear." The procedure word "roger" means "I have received your transmission correctly." It does not mean that the receiving operator agrees with the message, or will comply with it. Using standard procedure words, such as "over," "roger," and "out" is good operating practice.

When communications are difficult because of noise or weak signals, you can avoid confusion over unusual words by spelling them out using the standard phonetic alphabet. You should not shout into the microphone (or turn up the microphone to gain control) as this will distort your signal and make it less understandable, rather than louder. When received signals are weak, it might help to "open up" the squelch control. Normally, you would set this control just past the point where background noise is cut off. The volume control should be adjusted for a comfortable listening level.

Whenever your marine radio is on, but you are not using it to communicate, you must listen to the distress and calling frequency. You must maintain this watch instead of turning off your radio. If you want to guard (maintain a watch on) a different frequency, you may do so only if your equipment has capability for monitoring two channels simultaneously, or if you have an additional receiver to maintain the required watch on the distress and calling frequency. Voluntarily equipped vessels are not required to operate their radios, but if they are operated, they must comply with the above watch requirements.

The calling and distress frequencies for the maritime services are as follows:

- VHF (very high frequency): 156.8 MHz (Channel 16)
- MF (medium frequency): 2182 kHz

The U. S. Coast Guard and ships maintain a silent period on 2182 kHz for three minutes immediately after the hour and the half hour. During these silent periods, only distress or urgency messages may be transmitted.

Priority of communications

The order of priority for communications in the mobile service and the maritime mobile satellite service shall be as follows:

1. Distress calls, distress messages, and distress traffic indicate that a mobile station is threatened by grave and imminent danger and requests immediate assistance. Example: sinking ship.

2. Communications preceded by the urgency signal indicate that the calling station has a very urgent message to transmit concerning the safety of a ship, aircraft, or other vehicle, or the safety of a person. Example: man overboard, unable to stop or turn vessel.

3. Communications preceded by the safety signal indicate that the station is about to transmit a message concerning the safety of navigation or giving important meteorological warnings. Example: floating derelict in narrow navigational channel; hurricane warning, etc.

4. Communications relating to radio direction finding.

5. Communications relating to the navigation and safe movement of aircraft engaged in search and rescue operations.

6. Communications relating to the navigation, movements, and needs of ships, and weather observation messages destined for an official meteorological service.

7. Etatprioritenations are government radiotelegrams relative to the application of the United Nations Charter.

8. Etatpriorite are government radiotelegrams with priority and Government calls for which priority has been expressly requested.

9. Service communications relating to the working of the telecommunications service or to communications previously exchanged.

10. Government communications other than those shown in 7 and 8, ordinary private communications, RCT radiotelegrams, and press radiotelegrams.

Distress procedures

The radiotelephone alarm signal consists of two high-pitched alternating tones, each tone segment ¼ second in length. The tones continue for 30 to 60 seconds. If you should hear this alarm signal on a distress frequency or on any other marine channel, you should stop transmitting immediately and listen for any forthcoming distress messages.

The radiotelephone alarm signal shall only be used to announce:

1. That a distress call or message is about to follow;

2. The transmission of an urgent tornado or waterspout warning. In this case, the alarm signal may only be used by coast stations authorized by the Commission to do so; or

3. The loss of a person or persons overboard. In this case, the alarm signal may only be used when the assistance of other ships is required and cannot be satisfactorily obtained by the use of the urgency signal only. The alarm signal shall not be repeated by other stations. The message shall be preceded by the urgency signal. The frequency 2182 kHz is the international general calling and distress frequency for radiotelephony in the medium frequency band. It shall be used by ship, aircraft, and survival craft stations operating in the authorized bands between 1605 and 4000 kHz when requesting assistance from the maritime services. The frequency 156.8 MHz is the international distress, safety and calling frequency for radiotelephony for stations of the maritime mobile service when using frequencies in the authorized bands between 156 and 174 MHz.

In addition, these frequencies may be used for the transmission of:

1. The international urgency signal and very urgent messages,
2. The international safety signal and messages,
3. Brief radio operating signals, and
4. Brief test signals that might be necessary to determine whether the radio transmitting equipment of the station is in good working condition on this frequency.

The radiotelephone distress procedure shall consist of:

1. The radiotelephone alarm signal (whenever possible),
2. The distress call, and
3. The distress message.

The distress call sent by radiotelephony consists of:

1. The distress signal mayday, spoken three times,
2. The words "this is," followed by
3. The callsign (or name, if no callsign is assigned) of the mobile station in distress, spoken three times.

The distress message sent by radiotelephony consists of:

1. The distress call as specified above,
2. The name of the mobile station in distress,
3. Particulars of its position,
4. The nature of the distress,
5. The kind of assistance desired, and
6. Any other information that might facilitate rescue (for example, the length, color, and type of vessel, number of persons on board, etc.).

The acknowledgment of receipt of a distress message is transmitted, when radiotelephony is used, in the following form:

1. The callsign or other identification of the station sending the distress message, spoken three times,

2. The words "this iS,"

3. The callsign or other identification of the station acknowledging receipt of the message, spoken three times,

4. The words received, and

5. The distress signal mayday.

Every mobile station that acknowledges receipt of a distress message shall transmit, as soon as possible, the following information in the order shown:

1. Its name,
2. Its position, and
3. The speed at which it is proceeding towards, and the approximate time it will take to meet the mobile station in distress.

Before sending this message, the station shall assure that it will not interfere with the emissions of other stations better suited to render immediate assistance to the station in distress.

A mobile station or a land station that learns that a mobile station is in distress shall transmit a distress message in any of the following cases:

1. When the station in distress is not itself in a position to transmit the distress message;

2. When the master, or person responsible for the ship, aircraft, or vehicle in distress, or the person responsible for the land station considers that further help is necessary; and,

3. When, although not in a position to render assistance, it has heard a distress message which has not been acknowledged. When a mobile station transmits a distress message under these conditions, it shall take all necessary steps to notify the authorities who may be able to render assistance.

Distress traffic consists of all messages that relate to the immediate assistance required by a mobile station in distress. The control of distress traffic is the responsibility of the mobile station in distress, or of the station that has sent the distress message. These stations might, however, delegate control of the distress traffic to another station.

In radiotelephony, the urgency signal consists of the word PAN, spoken three times and transmitted before the call. The urgency signal and call, and the message following it, shall be sent on one of the international distress frequencies (2182 kHz or 156.8 MHz). However, stations that cannot transmit on a distress frequency may use any other available frequency on which attention might be attracted.

Mobile stations that hear the urgency signal shall continue to listen for at least three minutes. At the end of this period, if no urgency message has been heard, they may resume their normal service. However, land and mobile stations that are in communication on frequencies other than those used for the transmission of the urgency signal and of the call that follows it may continue their normal work without interruption, provided that the urgency message is not addressed "to all stations."

In radiotelephony, the safety signal consists of the word *security*, spoken three times and transmitted before the call. The safety signal and call shall be sent on one of the international distress frequencies (500 kHz radiotelegraph; 2182 kHz or 156.8 MHz radiotelephone). However, stations that cannot transmit on a distress frequency may use any other available frequency on which attention might be attracted.

Ship stations must use every precaution to ensure that, when conducting operational transmitter tests, the emissions of the station will not cause harmful interference. Radiation must be reduced to the lowest practicable value and, if feasible, should be entirely suppressed. When radiation is necessary or unavoidable, the testing procedure described in Section 80.101 of FCC Rules must be followed.

Each of the marine channels is reserved for a specific type of communication. You must use the correct channel for the type of communications you need. A complete list of VHF marine channels is contained in Section 80.371 of FCC Rules or in the Marine Radiotelephone Users Handbook.

Channel 22 is used by the U. S. Coast Guard to communicate with boaters. To contact the Coast Guard, you should first call them on Channel 16. Then follow the instructions that they give you. Channel 22 is limited to communications with the Coast Guard only. On the MF band, you should call the Coast Guard using 2182 kHz. The MF band frequency usually used for communicating with the Coast Guard is 2670 kHz. It is not to be used for other purposes.

The initial steps for performing a routine, on-the-air test of a marine radiotelephone transmitter are as follows:

- switch to a working frequency,
- listen carefully to ensure that the test transmissions will not be likely to interfere with other communications in progress, and,
- transmit the station callsign followed by the word "test" as a warning that test emissions are about to be made on that frequency.

If a VHF marine station that you are calling does not answer your first call, you may call that station a second time and a third time if necessary, waiting at least two minutes between each call. When the station doesn't reply to the third call, the calling must cease and not be renewed until after an interval of 15 minutes, unless there is no reason to believe that harmful interference will be caused to other communications in progress, in which case, the call sent three times at intervals of two minutes may be repeated after a pause of not less than three minutes. In the event of an emergency involving safety, the provisions of this paragraph shall not apply.

Ship stations transmitting on any authorized VHF bridge-to-bridge channel may be identified by the name of the ship in lieu of the callsign. Ship stations operating in a vessel traffic service system or on a waterway under the control of a U. S. Government agency or a foreign authority, when communicating with such an agency or authority may be identified by the name of the ship in lieu of the callsign, or as directed by the agency or foreign authority.

Ship radiotelephone logs

Logs of ships which are compulsorily equipped for radiotelephony must contain the following applicable log entries and the time of their occurrence:

1. A summary of all distress, urgency, and safety traffic;
2. A summary of communications between the ship station and the land or mobile stations;
3. A reference to important service incidents;
4. The position of the ship at least once a day;
5. The name of the operator at the beginning and end of the watch period;
6. The time the watch begins when the vessel leaves port, and the time it ends when the ship reaches port;
7. The time the watch is discontinued, including the reason, and the time the watch is resumed;
8. The times when storage batteries provided as a part of the required radiotelephone installation are placed on charge and taken off charge;
9. Results of required equipment tests, including specific gravity of lead-acid storage batteries and voltage readings of other types of batteries provided as a part of the compulsory equipment;
10. Results of inspections and tests of compulsorily fitted lifeboat radio equipment;
11. A daily statement about the condition of the required radiotelephone equipment, as determined by either normal communication or test communication.
12. When the master is notified about improperly operating radiotelephone equipment.

Further General Radiotelephone Operating Procedures are contained in FCC Rules Section 80.116.

Other operating reminders

To call another boat using VHF marine radio, you should use the following procedures:

1. As soon as communication is established by a call and reply, you should immediately change to the working frequency to leave the calling channel clear for other callers or possible emergency use.
2. Make certain that the working frequency you have selected is one which is permissible for use of the type of communications you wish to exchange.
3. It is generally a good idea to briefly monitor the working frequency you plan to use to make certain that it is not busy before you initiate a call on the

calling frequency. If it is known that the boat with which you wish to communicate maintains a dual watch on both the working frequency and the calling frequency, you should call on the working frequency. If you are uncertain, or if the vessel to be called does not have the capability for monitoring two channels at the same time, you should initiate the call on the calling frequency.

4. Keep your remarks brief and to the point. You must restrict your remarks to matters concerning the operation of the vessel or to business matters related to the operation of vessels. Except when you are making a radiotelephone call via a public coast station, discussions of personal affairs or other matters not pertaining to the safety, welfare or business of the operation of your vessel are not permitted.

Make certain that your radio is never left unattended in an exposed or unprotected position. When the operator is not present, the radio equipment must be secured so that it is not accessible to unauthorized persons who might board or be aboard your vessel.

Make certain that only persons who have been indoctrinated in proper operating procedures are permitted to operate the radio. Make certain that each person allowed to operate the radio possesses a Restricted Radiotelephone Operator Permit or a higher class of FCC radio operator license when required by FCC Rules for the type of vessel you have.

Do not use Channel 16 for other than call and reply under normal circumstances and safety, urgent and distress calls in emergency situations.

Do not use Channel 6 for routine communications with other vessels. It is reserved exclusively for safety communications. Use it for rendering assistance to disabled vessels, assisting the Coast Guard in search and rescue operations, and similar purposes.

Make certain that your radio is kept in good operating condition by having it serviced at reasonable intervals by a competent radio technician.

Protect the radio equipment from damage by water, moisture, physical abuse and electrical overloading. Never try to operate the equipment with the antenna disconnected. Keep the antenna and feedline in good condition.

Familiarize yourself with operational requirements of FCC Rules and Regulations.

Check your vessel's radio license from time to time. Apply for renewal before it expires.

Most vessels where the Marine Radio Operator Permit is required are also required to be inspected periodically by both the Coast Guard and FCC. Make certain that your required inspection certificates are posted and are not expired. Application for FCC inspection must be submitted at least 3 days in advance. Submissions of timely requests for inspection is the joint responsibility of the vessel's owner, operating agency, station licensee or the vessel's master. It is the duty of the radio operator, when not himself master, to bring an expiring inspection certificate to the attention of the vessel's master.

How to use your VHF marine radio (Information for Recreational Boaters)

General

Do I have to have a copy of the rules? Recreational boaters are not required to keep a copy of the FCC's rules. However, they are responsible for compliance with the FCC's rules. This bulletin is furnished for your information and guidance.

How to get a license

Do I need a license? You must have a ship station license before you use your radio. If you plan to dock in a foreign port, or are leaving a foreign port to dock in a U. S. port you must also have a Restricted Radiotelephone Operators Permit (RP) to operate a VHF Marine Radio. However, if you merely plan to sail in domestic or international waters without docking in any foreign ports, and your radio operates only VHF frequencies, you do not need an operators permit.

How do I apply for my license and for my RP?

- Use FCC Forms 506 to apply for a ship station license. The license term is for five years. You may not transfer this license to another person or boat.
- Use FCC Form 753 to apply for an RP, if required. There is no test required. The RP is issued for your lifetime.

May I operate my marine radio while my applications are being processed? You may operate your marine radio after you have mailed your application to the FCC, if:

- You fill out a temporary operating authority application (FCC Form 506A) and
- You keep this form with your station records. The completed form is your temporary operating authority.

This temporary operating authority is valid for 90 days after you mail your applications to the FCC.

How do I make changes during my license term? The following table tells you what you must do for changes during your license term:

Table 4-1
Requirements for Modifying Licenses

If you	You must
Change your mailing address	Tell the FCC in writing
Change your legal name	Tell the FCC in writing
Change the name of your boat	Tell the FCC in writing
Are a corporation and the ownership or control of the corporation changes	Apply for a new ship station license on FCC Form 506

Add or replace a transmitter which operates in the same frequency band	(No action required)
Add a transmitter that operates in a new frequency band	Apply for modification of your ship station license on FCC Form 506

Send your written notice of change to FCC, P. O. Box 1040, Gettysburg, PA 17326. Use FCC Form 506 to modify your ship station license.

How do I renew my license? Use FCC Form 405B to renew your license. The FCC will send you this form in approximately 120 days before your license expires. If you do not receive this form, you may use FCC Form 506 to renew your license.

If you send in your renewal form before your license expires, you may continue to operate under that license until the FCC acts on your application. You do not need a temporary permit, but you should keep a copy of the application you send the FCC.

You must stop transmitting as soon as your license expires, unless you have already sent your renewal application to the FCC.

What do I do if I lose my license or RP? If you lose your license, you must request a duplicate from the FCC, Gettysburg, PA 17326. Your request must include your name, vessel name, and your station callsign.

If you lose your RP, you must request a duplicate from the FCC, if required. Use FCC Form 753 to request a duplicate RP.

What must I do if I sell my boat? If you sell your boat, you must send your ship station license to the FCC, Gettysburg, PA 17326 for cancellation. You cannot transfer your ship station to another person or boat.

How to operate your radio

What type of equipment must I have? Your radio must be type accepted by the FCC. You can tell a type-accepted radio by the type acceptance label on the radio. You may look at a list of type-accepted radios at any FCC field office or at FCC headquarters.

The power output of your radio must not be more than 25 watts. You must also be able to lower the power of your radio to one watt or less.

Your radio must be able to transmit on Channel 16, Channel 6, and at least one other channel.

May I install and service my marine radio by myself? You may install your radio in your boat by yourself. However, all repairs or adjustments to your radio must be made by or under the supervision of an FCC-licensed general-class commercial operator. It is recommended that the radio be inspected by the service man when it is installed.

What channels may I use? Each channel is used only for certain types of messages. You must choose a channel that is available for the type of message you want to send.

Except where noted, channels are available for both ship-to-ship and ship-to-coast messages.

The channels listed in the table are the only channels you may use, even if your radio has more channels available.

How do I operate my marine radio?

- Maintain your watch   Whenever your radio is turned on (and not being used for messages), keep it tuned to Channel 16.
- Power   Try one watt first if the station being called is within a few miles. If no answer, you may switch to higher power.
- Calling coast stations   Call a coast station on its assigned channel. You may use Channel 16 when you do not know the assigned channel.
- Calling other boats or ships   Call other boats or ships on Channel 16. You may call on ship-to-ship channels when you know that the boat is listening on both a ship-to-ship channel and Channel 16. *Note:* To do this the boat has to have two separate receivers.
- Limits on calling   You must not call the same station for more than 30 seconds at a time. If you do not get a reply, wait at least two minutes before calling again. After three calling periods, wait at least 15 minutes before calling again.
- Change channels   After contacting another station on Channel 16, change immediately to a channel which is available for the type of message you want to send.
- Station identification   Identify your station by your FCC callsign at the beginning and end of each message. Identify in English.

What communications are prohibited? You must not transmit false distress or emergency messages. You must not transmit messages containing obscene, indecent, or profane words or meaning. You must not transmit general calls, signals, or messages, except in an emergency or if you are testing your radio (these are messages not addressed to a particular station), or when your boat is on land (for example, while the boat is on a trailer).

Do I have to keep a radio log? You do not have to keep a radio log.

Where do I get a copy of the rules? Write to:

Superintendent of Documents
Government Printing Office
Washington, DC 20402

Do I have to make my ship station available for inspection? Your station and your station records (station license and operator license or RP, if required) must be shown when requested by an authorized FCC representative.

What happens if I violate the rules? If it appears to the FCC that you have violated the Communications Act or the rules, the FCC might send you a written notice of the apparent violation.

If the violation notice covers a technical radio standard, you must stop using your radio. You must not use your radio until you have had all the technical problems fixed. You might have to have tests conducted. You might have to report the results of those test to the FCC. Test results must be signed by the commercial operator who conducted that test.

If the FCC finds that you have willfully or repeatedly violated the Communications Act or these rules, your license might be revoked and you might be fined or sent to prison.

How do I call another boat? Speak directly into the microphone in a normal tone of voice that is clear and distinct.

1. Make sure your radio is on.
2. Select Channel 16 (156.8 MHz) and listen to make sure it is not being used.
3. Press the microphone button and call the boat you wish to call. Say "(name of ship being called), this is (your ship's name and callsign)."
4. Once contact is made on Channel 16, you must switch to a ship-to-ship channel.
5. After communications is completed, each ship must give its callsign and switch to Channel 16.

How do I place a call through a public coast station? Speak directly into the microphone in a normal tone of voice that is clear and distinct.

1. Make sure your radio is on.
2. Select correct channel for the public coast station and listen to make sure it is not being used.
3. Press microphone button and say "(Name of coast station), this is (Your callsign)."
4. When coast station operator answers say, "this is (Name of boat, callsign and billing number if assigned). Placing a call to (city, telephone number desired)," inform operator of type billing desired.
5. After completion of call, say "(Name of boat, callsign) out."

What are the marine emergency signals? The three spoken international emergency signals are:

- Mayday   The distress signal mayday is used to indicate that a station is threatened by grave and imminent danger and requests immediate assistance. Mayday has priority over all other messages.
- Pan pan   The urgency signal pan pan is used when the safety of the vessel or person is in jeopardy.
- Security   The safety signal security is used for messages about the safety of navigation or important weather warnings.

When using an international emergency signal, the appropriate signal is to be spoken three times and transmitted before the call.

You must give any message beginning with one of these signals priority over routine messages.

What is the marine distress procedure? Marine Distress Communications Form (speak slowly, clearly, and calmly).

1. Make sure your radio is on.
2. Select VHF Channel 16 (156.8 MHz).
3. Press microphone button and say: "Mayday   Mayday   Mayday."
4. Say "This is _____." (your boat name/callsign repeated three times).
5. Say "Mayday _____." (your boat name).
6. Tell where you are (what navigational aids or landmarks are near).
7. State the nature of your distress.
8. Give number of persons aboard and conditions of any injured.
9. Estimate present seaworthiness of your boat.
10. Briefly describe your boat _____ Feet. (Length) _____, _____ Hull, _____ (Type) (Color)
11. Say: "I will be listening on Channel 16."
12. End message by saying, "this is (your boat name and callsign), over."
13. Release microphone button and listen. Someone should answer. If they do not, repeat call, beginning at item 3 above.

Table 4-2
Permitted Channel Usage

Type of message	Channel(s) suitable
Distress safety and calling Use this channel to get the attention of another station (calling) or in emergencies (distress and safety).	16
Intership safety Use this channel for ship-to-ship safety messages and for search and rescue messages and ships and aircraft of the Coast Guard.	6
Coast Guard liaison Use this channel to talk to the Coast Guard (but first make contact on Channel 16).	22
Noncommercial Working channels for recreational boats. Messages must be about the needs of the vessel. Typical uses include fishing reports, rendezvous, scheduling repairs and berthing information. Use Channel 72 only for ship-to-ship messages.	9, 68, 69, 71, 72, 78
Commercial Working channels for working vessels only. Messages must be about business or the needs of the vessel. Use Channels 8, 67, and 88 only for ship-to-ship messages.	1, 7, 8, 9, 10, 11, 18, 19, 63, 67, 79, 80, 88[1]

Public correspondence (marine operator) Use these channels to call the marine operator at a public coast station. By contacting a public coast station, you can make and receive calls from telephones on shore. Except for distress calls, public coast stations usually charge for this service.	24, 25, 26, 27, 28, 84, 85, 86, 87, 88[2]
Port operations These channels are used in directing the movement of ships in or near ports, locks, or waterways. Messages must be about the operational handling movement and safety of ships. In certain major ports Channels 11, 12, and 14 are being used for the Vessel Traffic Service systems being developed by the Coast Guard and are not available for general port operations messages. Channel 77 is limited to intership communications to and from pilots.	1, 5[3], 12, 14, 20, 63, 65, 66, 73, 74, 77
Navigational (Also known as the bridge-to-bridge channel.) This channel is available to all ships. Messages must be about vessel navigation, for example, passing or meeting other vessels. You must keep your messages short. Your power output must not be more than 1 watt. This is also the main working channel at most locks and drawbridges.	13, 67
Maritime control This channel may be used to talk to ships and coast stations operated by State or local governments. Messages must pertain to regulation and control, boating activities, or assistance to ships.	17
Digital selective calling Use this channel for distress and safety calling and for general purpose calling using only digital selective calling techniques.	70
Weather On these channels you may receive weather broadcasts of the National Oceanic and Atmospheric Administration. These channels are only for receiving. You cannot transmit on them.	WX-1 (162.550 MHz), WX-2 (162.400 MHz), WX-3 (162.475 MHz)

[1] Not available in the Great Lakes, the St. Lawrence Seaway, or the Puget Sound and the Strait of Juan de Fuca and its approaches.

[2] Only for use in the Great Lakes, St. Lawrence Seaway, and Puget Sound and the Strait Juan de Fuca and its approaches.

[3] Available only in the Houston and New Orleans areas.

5
CHAPTER

Element 1
Laws and operating practices

If you are planning to take an FCC test, consider this section to be required material. The questions in this chapter were also given in the book titled *The TAB Book for Communications Licensing and Certification Examinations*. However, the questions and answers have been scrambled. That gives you a new experience in answering questions on this subject. The answers are at the end of the chapter. You should review this material periodically.

1. What is not a good soldering practice in electronic circuits?
   A. Be certain parts do not move while solder is cooling.
   B. Prevent corrosion by never using flux.
   C. Clean parts sufficiently.
   D. Use adequate heat.

2. An auto alarm signal consists of two sine-wave audio tones transmitted alternately at what frequencies?
   A. None of these.
   B. 1300 and 220 kHz.
   C. 500 and 100 kHz.
   D. 121.5 and 243 MHz.

3. 6:00 PM PST is equal to what time in UTC?
   A. 1300.
   B. 2300.
   C. 1800.
   D. 0200.

4. 100 statute miles equals how many nautical miles?

    A. 13.
    B. 173.
    C. 108.
    D. 87.

5. Which of the following is an acceptable method of solder removal from holes in a printed board?

    A. Power drill.
    B. Soldering iron and a suction device.
    C. Toothpick.
    D. Compressed air.

6. 10 statute miles per hour equals how many knots?

    A. 3.
    B. 5.
    C. 8.7.
    D. 11.5.

7. 3:00 PM Central Standard Time is:

    A. 0300 UTC.
    B. 1800 UTC.
    C. 2100 UTC.
    D. 100 UTC.

8. The ILS localizer measures what deviation of an aircraft?

    A. Distance between aircraft.
    B. Ground speed.
    C. Vertical.
    D. Horizontal.

9. The condition of a storage battery is determined with a:

    A. Hydrometer.
    B. FET.
    C. Manometer.
    D. Hygrometer.

10. Which of these will be useful for insulation at UHF?

    A. Lead.
    B. Wax-impregnated paper.
    C. Mica.
    D. Rubber.

11. What is the purpose of flux?

A. Both C and D.
B. Acid cleans printed circuit connections.
C. Prevents oxidation during soldering.
D. Removes oxides from surfaces to be joined.

12. 2.3 statute miles equals how many nautical miles?

    A. 1.
    B. 1.73.
    C. 1.5.
    D. 2.

13. When soldering electronic circuits be sure to:

    A. Use minimum solder.
    B. Heat wires until sweating begins.
    C. Use maximum heat.
    D. Use sufficient heat.

14. One statute mile equals how many nautical miles?

    A. 0.7.
    B. 0.87.
    C. 1.5.
    D. 3.8.

15. 2300 UTC time is:

    A. 6 AM EST.
    B. 10 AM EST.
    C. 3 PM PST.
    D. 2 PM CST.

16. Waveguides are not utilized at VHF or UHF frequencies because:

    A. Grounding problems.
    B. Large dimensions of waveguides are not practical.
    C. Resistance to high-frequency waves.
    D. Characteristic impedance.

17. What is a descending pass for a satellite?

    A. A pass from south to north.
    B. A pass from east to west.
    C. A pass from west to east.
    D. A pass from north to south.

18. What is an ascending pass for a satellite?

    A. A pass from north to south.
    B. A pass from south to north.

C. A pass from east to west.

D. A pass from west to east.

19. In facsimile, how are variations in picture brightness and darkness converted into voltage variations?

A. With an optoisolator.

B. With a photodetector.

C. With a Hall-effect transistor.

D. With an LED.

20. What is the term for the transmission of printed pictures by radio for the purpose of a permanent display?

A. ACSSB.

B. Xerography.

C. Facsimile.

D. Television.

21. What is the approximate transmission time for a facsimile picture transmitted by a radio station?

A. ⅟₆₀ second per frame at 240 lines per minute.

B. Approximately 6 seconds per frame at 240 lines per minute.

C. Approximately 3.3 minutes per frame at 240 lines per minute.

D. Approximately 6 minutes per frame at 240 lines per minute.

22. What is the modern standard scan rate for a facsimile picture transmitted by a radio station?

A. The modern standard is 60 lines per second.

B. The modern standard is 150 lines per second.

C. The modern standard is 50 lines per minute.

D. The modern standard is 240 lines per minute.

23. What is facsimile?

A. The transmission of printed pictures for permanent display on paper.

B. The transmission of video by television.

C. The transmission of still pictures by slow-scan television.

D. The transmission of characters by radioteletype that form a picture when printed.

24. Why does the received signal from a satellite stabilized by a computer-pulsed electromagnet exhibit a fairly rapid pulsed fading effect?

A. Because of the Doppler effect.

B. Because of the satellite's low orbital altitude.

C. Because of ionospheric absorption.

D. Because the satellite is rotating.

25. Why does the downlink frequency appear to vary by several kHz during a low-earth-orbit satellite pass?

   A. The distance between the satellite and ground station is changing, causing the Doppler effect.
   B. The distance between the satellite and ground station is changing, causing the Boyle's law effect.
   C. The distance between the satellite and ground station is changing, causing the Bernoulli effect.
   D. The distance between the satellite and ground station is changing, causing the Kepler effect.

26. What are the two basic types of linear transponders used in satellites?

   A. Amplitude modulated and frequency modulated.
   B. Phase 2 and Phase 3.
   C. Geostationary and elliptical.
   D. Inverting and noninverting.

27. What is a linear transponder?

   A. A device used to change an FM emission to an AM emission.
   B. An amplifier for SSB transmissions.
   C. A device that receives and retransmits signals of any mode in a certain pass-band.
   D. A repeater that passes only linear or binary signals.

28. What is the period of a satellite?

   A. The time it takes a satellite to travel from perigee to apogee.
   B. The amount of time it takes for a satellite to complete one orbit.
   C. The point on an orbit where satellite height is minimum.
   D. An orbital arc that extends from 60 degrees.

29. A room temperature of +30.0 degrees Celsius is equivalent to how many degrees Fahrenheit?

   A. 86.
   B. 95.
   C. 83.
   D. 104.

30. What is the frequency range of UHF?

   A. 30 to 300 MHz.
   B. 3 to 30 kHz.
   C. 0.3 to 3 MHz.
   D. 0.3 to 3 GHz.

31. When is the silent period on 2182 kHz, when only emergency communications may occur?

    A. Three minutes at the beginning of every hour and half hour.
    B. No designated period; silence is maintained only when a distress call is received.
    C. At all times.
    D. One minute at the beginning of every hour and half hour.

32. When and how may Class A and B EPIRBs be tested?

    A. At any time ship is at sea.
    B. Within first 1 minute of hour, test not to exceed 1 minute.
    C. Within first 3 minutes of hour; tests not to exceed 30 seconds.
    D. Within the first 5 minutes of the hour; tests not to exceed 3 audible sweeps or one second, whichever is longer.

33. What has most priority:

    A. Security.
    B. Safety.
    C. Distress.
    D. Urgent.

34. What is the international radiotelephone distress call?

    A. The alternating two tone signal produced by the radiotelephone alarm signal generator.
    B. For radiotelephone use, any words or message which will attract attention may be used.
    C. "Mayday, mayday, mayday; This is . . .;" followed by the callsign (or name, if no callsign assigned) of the mobile station in distress, spoken three times.
    D. "SOS, SOS, SOS; This is . . .;" followed by the callsign of the station (repeated 3 times).

35. In the International Phonetic Alphabet, the letter E, M, and S are represented by the words:

    A. Element, Mister, Scooter.
    B. Echo, Mike, Sierra.
    C. Equator, Mike, Sonar.
    D. Echo, Michigan, Sonar.

36. Frequencies which have substantially straight-line propagation characteristics similar to that of light waves are:

    A. Frequencies above 50,000 kHz.
    B. Frequencies between 1000 kHz and 3000 kHz.
    C. Frequencies between 500 kHz and 1000 kHz.
    D. Frequencies below 500 kHz.

37. Atmospheric noise or static is not a great problem:

    A. At frequencies above 30 MHz.
    B. At frequencies above 1 MHz.
    C. At frequencies below 5 MHz.
    D. At frequencies below 20 MHz.

38. What is an urgency transmission?

    A. A communications transmission concerning the safety of a ship, aircraft or other vehicle, or of some person on board or within sight.
    B. A communications alert that important personal messages must be transmitted.
    C. Health and welfare traffic which impacts the protection of on-board personnel.
    D. A radio distress transmission affecting the security of humans or property.

39. What is a maritime mobile repeater station?

    A. A one way low-power communications system used in the maneuvering of vessels.
    B. A mobile radio station that links two or more public coast stations.
    C. An automatic on-board radio station that facilitates the transmissions of safety communications aboard ship.
    D. A fixed land station used to extend the communications range of ship and coast stations.

40. What is distress traffic?

    A. All messages relative to the immediate assistance required by a ship, aircraft or other vehicle in imminent danger.
    B. Internationally recognized communications relating to emergency situations.
    C. Health and welfare messages concerning the immediate protection of property and safety of human life.
    D. In radiotelegraphy, SOS sent as a single character; in radiotelephony, the speaking of the word, "Mayday."

41. What is a requirement of every commercial operator on duty and in charge of a transmitting system?

    A. A copy of the operator's license must be supplied to the radio station's supervisor as evidence of technical qualification.
    B. The FCC Form 756 certifying the operator's qualifications must be readily available at the transmitting system site.
    C. The original license or a photocopy must be posted or in the operator's personal possession and available for inspection.
    D. A copy of the Proof-of-Passing Certificate (PPC) must be on display at the transmitter location.

42. Who is required to make entries on a required service or maintenance log?

    A. The technician who actually makes the adjustments to the equipment.
    B. Any commercial radio operator holding at least a Restricted Radiotelephone Operator Permit.
    C. The operator responsible for the station operation or maintenance.
    D. The licensed operator or a person whom he or she designates.

43. Which of the following persons are ineligible to be issued a commercial radio operator license?

    A. U.S. Military radio operators who are still on active duty.
    B. Foreign maritime radio operators unless they are certified by the International Maritime organization (IMO).
    C. Handicapped persons with uncorrected disabilities which affect their ability to perform all duties required of commercial radio operators.
    D. Individuals who are unable to send and receive correctly by telephone spoken messages in English.

44. What authority does the Marine Radio Operator Permit confer?

    A. The nontransferable right to install, operate and maintain any type-accepted radio transmitter.
    B. Confers authority to operate licensed radio stations in the Aviation, Marine and International Fixed Public Radio Services.
    C. Allows the radio operator to maintain equipment in the Business Radio Service.
    D. Grants authority to operate commercial broadcast stations and repair associated equipment.

45. What is the Global Maritime Distress and Safety System (GMDSS)?

    A. The international organization charged with the safety of ocean-going vessels.
    B. An association of radio officers trained in emergency procedures.
    C. An emergency radio service employing analog and manual safety apparatus.
    D. An automated ship-to-shore distress alerting systems using satellite and advanced terrestrial communications systems.

46. What is the standard video level, in percent PEV, for blanking?

    A. 100%.
    B. 75%.
    C. 12.5%.
    D. 0%.

47. What is the standard video level, in percent PEV, for white?

    A. 100%.

B. 70%.
C. 12.5%.
D. 0%.

48. What is the standard video level, in percent PEV, for black?

    A. 100%.
    B. 70%.
    C. 12.5%.
    D. 0%.

49. What is the standard video voltage level between the sync tip and the whitest white at TV camera outputs and modulator inputs?

    A. 5 V RMS.
    B. 12 Vdc.
    C. 120 IEEE units.
    D. 1 V peak-to-peak.

50. What is blanking in a video signal?

    A. Transmitting a black-and-white test pattern.
    B. Turning off the scanning beam at the conclusion of a transmission.
    C. Turning off the scanning beam while it is traveling from right to left and from bottom to top.
    D. Synchronization of the horizontal and vertical sync-pulses.

51. What type of antenna can be used to minimize the effects of spin modulation and Faraday rotation?

    A. A log-periodic dipole array.
    B. An isotropic antenna.
    C. A circularly polarized antenna.
    D. A nonpolarized antenna.

52. What ferrite device can be used instead of a duplexer to isolate a microwave transmitter and receiver when both are connected to the same antenna?

    A. Simplex.
    B. Magnetron.
    C. Circulator.
    D. Isolator.

53. Solder is:

    A. 70% lead 30% tin.
    B. 60% lead 40% tin.
    C. 40% lead 60% tin.
    D. 50% lead 50% tin.

54. 1.73 nautical miles equals how many statute miles?

    A. 1.

    B. 1.73.

    C. 1.5.

    D. 2.

55. One nautical mile is equal to how many statute miles?

    A. 1.15.

    B. 1.73.

    C. 8.3.

    D. 1.5.

56. If you are listening to an FM radio station at 100.6 MHz on a car radio when an airplane in the vicinity is transmitting at 121.2 MHz and your car radio receives interference the possible problem could be:

    A. Intermodulation or coupling.

    B. Image frequency.

    C. Poor Q receiver.

    D. Improper shielding in receive.

57. A 25-MHz amplitude-modulated transmitter's actual carrier frequency is 25.00025 MHz without modulation and is 24.99950 MHz when modulated. What statement is true?

    A. If the authorized frequency tolerance is 0.005% for the 25 MHz band this transmitter is operating legally.

    B. Modulation should not change carrier frequency.

    C. If the allowed frequency tolerance is 0.002% this is an illegal transmission.

    D. If the allowed frequency tolerance is 0.001% this is an illegal transmission.

58. Licensed radiotelephone operators are not required on board ships for:

    A. Any of the below.

    B. Installation of a VHF transmitter in a ship station where the work is performed by or under the immediate supervision of the licensee of the ship station.

    C. Ship radar, provided the equipment is nontunable, pulse type Magnetron and can be operated by means of exclusively external controls.

    D. Voluntarily equipped ship stations on domestic voyages operating on VHF channels.

59. The auto alarm device for generating signals shall be:

    A. None of the below.

    B. Tested weekly using a dummy load.

    C. Tested every three months using a dummy load.

    D. Tested monthly using a dummy load.

60. How long should station logs be retained when there are no entries relating to distress or disaster situations?

    A. For a period of one year from the date of entry.
    B. Indefinitely, or until destruction is specifically authorized by the U.S. Coast Guard.
    C. Until authorized by the Commission in writing to destroy them.
    D. For a period of three years from the date of entry unless notified by the FCC.

61. The system of substituting words for corresponding letters is called:

    A. 10 codes.
    B. Mnemonic system.
    C. Phonetic system.
    D. International code system.

62. VHF ship station transmitters must have the capability of reducing carrier power to:

    A. 50 W.
    B. 25 W.
    C. 10 W.
    D. 1 W.

63. Which VHF channel is used only for digital selective calling?

    A. Channel 6.
    B. Channel 22A.
    C. Channel 16.
    D. Channel 70.

64. When testing is conducted on 2182 kHz or 156.8 MHz testing should not continue for more than in any 5 minute period.

    A. None of the below.
    B. 2 minutes.
    C. 1 minute.
    D. 10 seconds.

65. What channel must compulsorily equipped vessels monitor at all times in the open sea?

    A. Channel 6, 156.3 MHz.
    B. Channel 22A, 157.1 MHz.
    C. Channel 16, 156.8 MHz.
    D. Channel 8, 156.4 MHz.

66. The FCC may suspend an operator license upon proof that the operator:

    A. Any of the below.

B. Has transmitted obscene language.

C. Has willfully damaged transmitter equipment.

D. Has assisted another to obtain a license by fraudulent means.

67. When a message has been received and will be complied with say:

A. Wilco.

B. Roger.

C. Over.

D. Mayday.

68. When attempting to contact other vessels on Channel 16:

A. Both C and D.

B. Channel 16 is used for emergency calls only.

C. If no answer is received, wait 2 minutes before calling vessel again.

D. Limit calling to 30 seconds.

69. When your transmission is ended and you expect no response, say:

A. Clear.

B. Roger.

C. Over.

D. Break.

70. What safety signal call word is spoken three times, followed by the station call letters spoken three times, to announce a storm warning, danger to navigation, or special aid to navigation?

A. Safety.

B. Security.

C. Mayday.

D. Pan.

71. Survival craft EPIRBs are tested:

A. All of the below.

B. With radiation reduced to a level not to exceed 25 microvolts per meter.

C. With a dummy load having the equivalent impedance of the antenna affixed to the EPIRB.

D. With a manually activated test switch.

72. What is the authorized frequency for an on-board ship repeater for use with a mobile transmitter operating at 467.750 MHz?

A. 467.825 MHz.

B. 467.800 MHz.

C. 467.775 MHz.

D. 457.525 MHz.

73. What do you do if the transmitter aboard your ship is operating off-frequency, overmodulating or distorting?

    A. Make a notation in station operating log.
    B. Reduce audio volume level.
    C. Stop transmitting.
    D. Reduce to low power.

74. What is the required daytime range of a radiotelephone station aboard a 900-ton ocean-going cargo vessel?

    A. 500 miles.
    B. 150 miles.
    C. 50 miles.
    D. 25 miles.

75. When may you test a radiotelephone transmitter on the air?

    A. After reducing transmitter power to 1 W.
    B. At any time as necessary to assure proper operation.
    C. Only when authorized by the Commission.
    D. Between midnight and 6:00 AM local time.

76. Each cargo ship of the United States which is equipped with a radiotelephone station for compliance with Part II of Title III of the Communications Act shall while being navigated outside of a harbor or port keep a continuous watch on:

    A. Cargo ships are exempt from radio watch regulations.
    B. Both C and D.
    C. 156.8 MHz.
    D. 2182 kHz.

77. Tests of survival craft radio equipment, except EPIRBs and two-way radiotelephone equipment, must be conducted:

    A. When required by the Commission.
    B. Both C and D below.
    C. Within 24 hours prior to departure when a test has not been conducted within a week of departure.
    D. At weekly intervals while the ship is at sea.

78. If a ship radio transmitter signal becomes distorted:

    A. Reduce audio amplitude.
    B. Use minimum modulation.
    C. Reduce transmitter power.
    D. Cease operations.

79. What is the radiotelephony calling and distress frequency?

    A. 2182R2647.
    B. 2182 kHz.
    C. 500R122JA.
    D. 500 kHz.

80. An operator or maintainer must hold a General Radiotelephone Operator License to:

    A. All of the below.
    B. Operate radiotelephone equipment with more than 1500 W of peak envelope power on cargo ships over 300 gross tons.
    C. Operate voluntarily equipped ship maritime mobile or aircraft transmitters with more than 1000 W of peak envelope power.
    D. Adjust or repair FCC licensed transmitters in the aviation, maritime and international fixed public radio services.

81. Overmodulation is often caused by:

    A. Shouting into microphone.
    B. Weather conditions.
    C. Station frequency drift.
    D. Turning down audio gain control.

82. When pausing briefly for station copying message to acknowledge, say:

    A. Stop.
    B. Wilco.
    C. Over.
    D. Break.

83. What is a good practice when speaking into a microphone in a noisy location?

    A. Shield microphone with hands.
    B. Increase monitor audio gain.
    C. Change phase in audio circuits.
    D. Overmodulation.

84. Marine transmitters should be modulated between:

    A. 75% to 120%.
    B. 85% to 100%.
    C. 70% to 105%.
    D. 75% to 100%.

85. The master or owner of a vessel must apply how many days in advance for an FCC ship inspection?

    A. 24 hours.

B. 3 days.

C. 30 days.

D. 60 days.

86. What emergency radio testing is required for cargo ships?

    A. All of the below.

    B. Specific gravity check in lead acid batteries, or voltage under load for dry cell batteries.

    C. Full power carrier tests into dummy load.

    D. Tests must be conducted weekly while ship is at sea.

87. When testing is conducted within the 2170 to 2194 kHz and 156.75 to 156.85 MHz bands, transmissions should not continue for more than _____ in any 15-minute period.

    A. No limitation.

    B. 5 minutes.

    C. 1 minute.

    D. 15 seconds.

88. What channel must VHF-FM equipped vessels monitor at all times the station is operated?

    A. Channel 1A; 156.07 MHz.

    B. Channel 5A; 156.25 MHz.

    C. Channel 16; 156.8 MHz.

    D. Channel 8; 156.4 MHz.

89. In the FCC rules the frequency band from 30 to 300 MHz is also known as:

    A. High Frequency (HF).

    B. Medium Frequency (MF).

    C. Ultra High Frequency (UHF).

    D. Very High Frequency (VHF).

90. Frequencies used for portable communications on board ship:

    A. 457.525 to 467.825 MHz.

    B. 2900 to 3100 MHz.

    C. 1636.5 to 1644 MHz.

    D. 9300 to 9500 MHz.

91. A reserve power source must be able to power all radio equipment plus an emergency light system for how long?

    A. 6 hours.

    B. 8 hours.

    C. 12 hours.

    D. 24 hours.

92. One nautical mile is approximately equal to how many statute miles?

    A. 1.47 statute miles.

    B. 1.15 statute miles.

    C. 1.83 statute miles.

    D. 1.61 statute miles.

93. How often is the auto alarm tested?

    A. Each day on 2182 kHz using a dummy antenna.

    B. Weekly on frequencies other than the 2182 kHz distress frequency using a dummy antenna.

    C. Monthly on 121.5 MHz using a dummy load.

    D. During the 5-minute silent period.

94. When should both the callsign and the name of the ship be mentioned during radiotelephone transmissions?

    A. Within 100 miles of any shore.

    B. When transmitting on 2182 kHz.

    C. During an emergency.

    D. At all times.

95. Which of the following transmissions are not authorized in the maritime service?

    A. Transmissions to test the operating performance of on-board station equipment.

    B. Needless or superfluous radiocommunications.

    C. Message handling on behalf of third parties for which a charge is rendered.

    D. Communications from vessels in dry dock undergoing repairs.

96. What regulations govern the use and operation of FCC-licensed ship stations operating in international waters?

    A. Those of the FCC's Aviation and Marine Branch, PRB, Washington, DC 20554.

    B. The Maritime Mobile Directives of the International Telecommunication Union.

    C. Part 80 of the FCC Rules plus the International Radio Regulations and agreements to which the United States is a party.

    D. The regulations of the International Maritime Organization (IMO) and Radio Officers Union.

97. What are the antenna requirements of a VHF telephony coast, marine utility, or ship station?

    A. The antenna must be capable of being energized by an output in excess of 100 W.

    B. The horizontally polarized antenna must be positioned so as not to cause excessive interference to other stations.

C. The antenna array must be type accepted for 30- to 200-MHz operation by the FCC.

D. The shore or on-board antenna must be vertically polarized.

98. Where do you submit an application for inspection of a ship radio station?

    A. To the nearest International Maritime Organization (IMO) review facility.
    B. To the Engineer-in-Charge of the FCC District Office nearest the proposed place of inspection.
    C. To the Federal Communications Commission, Washington, DC 20554.
    D. To a Commercial Operator Licensing Examination Manager (COLE Manager).

99. What is a requirement of all marine transmitting apparatus used aboard United States vessels?

    A. Programming of all maritime channels must be performed by a licensed Marine Radio Operator.
    B. Certification is required by the International Maritime Organization (IMO).
    C. Equipment must be approved by the U.S. Coast Guard for maritime mobile use.
    D. Only equipment that has been type accepted by the FCC for Part 80 operations is authorized.

100. What is a safety transmission?

    A. A radiotelegraphy alert preceded by the letters "XXX" sent three times.
    B. A communications transmission which indicates that a station is preparing to transmit an important navigation or weather warning.
    C. Health and welfare traffic concerning the protection of human life.
    D. A radiotelephony warning preceded by the words "pan."

101. What is the internationally recognized urgency signal?

    A. The pronouncement of the word "mayday."
    B. The word "pan" spoken three times before the urgent call.
    C. Three oral repetitions of the word "safety" sent before the call.
    D. The letters "TTT" transmitted three times by radiotelegraphy.

102. What is a ship earth station?

    A. An automated ship-to-shore distress alerting system.
    B. A communications system that provides line-of-sight communications between vessels at sea and coast stations.
    C. A mobile satellite station located on board a vessel.
    D. A maritime mobile-satellite station located at a coast station.

103. Loran C operates in what frequency band?

    A. LF; 30 to 300 kHz.

    B. MF; 300 to 3000 kHz.
    C. HF; 3 to 30 MHz.
    D. VHF; 30 to 300 MHz.

104. Shipboard transmitters using F3E emission (FM voice) may not exceed what carrier power?

    A. 25 W.
    B. 100 W.
    C. 250 W.
    D. 500 W.

105. Omega operates in what frequency band?

    A. 300 to 3000 kHz.
    B. 30 to 300 kHz.
    C. 3 to 30 kHz.
    D. Below 3 kHz.

106. The HF (high frequency) band is:

    A. 300 to 3000 MHz.
    B. 30 to 300 MHz.
    C. 3 to 30 GHz.
    D. 3 to 30 MHz.

107. Portable ship units, handhelds, or walkie-talkies used as an associated ship unit:

    A. All of the below.
    B. Must not transmit from shore or to other vessels.
    C. May communicate only with the mother ship and other portable units and small boats belonging to mother ship.
    D. Must operate with 1 W and be able to transmit on Channel 16.

108. What is the second in order of priority?

    A. Mayday.
    B. Safety.
    C. Distress.
    D. Urgent.

109. When a ship is sold:

    A. Continue to operate; license automatically transfers with ownership.
    B. The old license is valid until it expires.
    C. FCC inspection of equipment is required.
    D. The new owner must apply for a new license.

110. Two-way communications with both stations operating on the same frequency is:

    A. Multiplex.

    B. Simplex.

    C. Duplex.

    D. Radiotelephone.

111. Which of the following represent the first three letters of the phonetic alphabet?

    A. Adam Brown Chuck.

    B. Alpha Baker Crystal.

    C. Adam Baker Charlie.

    D. Alpha Bravo Charlie.

112. A marine public coast station operator may not charge a fee for what type of communication?

    A. All of the below.

    B. Distress.

    C. Storm updates.

    D. Port Authority transmissions.

113. That maritime MF radiotelephone silence periods begin at _____ and _____ minutes past the UTC hour.

    A. :05, :35.

    B. :20, :40.

    C. :00, :30.

    D. :15, :45.

114. What information must be included in a distress message?

    A. All of the below.

    B. Type of distress and specifics of help requested.

    C. Location.

    D. Name of vessel.

115. When all of a transmission has been received, say:

    A. Wilco.

    B. Received.

    C. Roger.

    D. Attention.

116. To indicate a response is expected, say:

    A. Break.

    B. Over.

    C. Roger.

    D. Wilco.

117. As an alternative to keeping watch on a working frequency in the band 1600 to 4000 kHz, an operator must tune station receiver to monitor 2182 kHz:
    A. During the silence periods each hour.
    B. During daytime hours of service.
    C. During distress calls only.
    D. At all times.

118. If your transmitter is producing spurious harmonics or is operating at a deviation from the technical requirements of the station authorization:
    A. Reduce power immediately.
    B. Cease transmission.
    C. Repair problem within 24 hours.
    D. Continue operating until returning to port.

119. What is the procedure for testing a 2182-kHz ship radiotelephone transmitter with full carrier power while out at sea?
    A. It is not permitted to test on the air.
    B. Simply say: "This is (call letters) testing." If all meters indicate normal values, it is assumed transmitter is operating properly.
    C. Switch transmitter to another frequency before testing.
    D. Reduce to low power, then transmit test tone.

120. Each cargo ship of the United States that is equipped with a radiotelephone station for compliance with the Safety Convention shall, while at sea:
    A. Keep continuous watch on 2182 kHz using a watch receiver that has a loudspeaker and auto alarm distress frequency watch receiver.
    B. Reduce peak envelope power on 156.8 MHz during emergencies.
    C. Keep the radiotelephone transmitter operating at full 100% carrier power for maximum reception on 2182 kHz.
    D. Not transmit on 2182 kHz during emergency conditions.

121. How should the 2182-kHz auto alarm be tested?
    A. On 2182 kHz into antenna.
    B. On 2182 kHz into dummy load.
    C. On a different frequency into dummy load.
    D. On a different frequency into antenna.

122. International laws and regulations require a silent period on 2182 kHz:
    A. Both C and D below.
    B. For the first minute of every quarter-hour.
    C. For three minutes immediately after the half-hour.
    D. For three minutes immediately after the hour.

123. Survival craft emergency transmitter tests may not be made:

    A. All of the below.

    B. Within 5 minutes of a previous test.

    C. Without using station callsign, followed by the word "test."

    D. For more than 10 seconds.

124. The urgency signal concerning the safety of a ship, aircraft, or person shall be sent only on the authority of:

    A. An FCC-licensed operator.

    B. Either C or D below.

    C. Person responsible for mobile station.

    D. Master of ship.

125. Maritime emergency radios should be tested:

    A. Both C and D.

    B. Every 24 hours.

    C. Weekly while the ship is at sea.

    D. Before each voyage.

126. Identify a ship station's radiotelephone transmissions by:

    A. Both B and C.

    B. Name of the vessel.

    C. Callsign.

    D. Country of registration.

127. What should an operator do to prevent interference?

    A. Both B and C.

    B. Transmissions should be as brief as possible.

    C. Monitor channel before transmitting.

    D. Turn off transmitter when not in use.

128. Under what license are handheld transceivers covered when used on board a ship at sea?

    A. No license is needed.

    B. Walkie-talkie radios are illegal to use at sea.

    C. Under the authority of the licensed operator.

    D. The ship station license.

129. What is the general obligation of a coast or marine-utility station?

    A. To broadcast warnings and other information for the general benefit of all mariners.

    B. To transmit lists of callsigns of all fixed and mobile stations for which they have traffic.

    C. To acknowledge and receive all calls directed to it by ship or aircraft stations.

D. To accept and dispatch messages without charge that are necessary for the business and operational needs of ships.

130. When does a maritime radar transmitter identify its station?

    A. By a transmitter identification label (TIL) secured to the transmitter.
    B. Radar transmitters must not transmit station identification.
    C. At 20-minute intervals using an automatic transmitter identification system.
    D. By radiotelegraphy at the onset and termination of operation.

131. Ordinarily, how often would a station using a telephony emission identify?

    A. At 20-minute intervals.
    B. At the beginning and end of each communication and at 15-minute intervals.
    C. At 15-minute intervals unless public correspondence is in progress.
    D. At least every 10 minutes.

132. Who is responsible for payment of all charges accruing to other facilities for the handling or forwarding of messages?

    A. The licensed commercial radio operator transmitting the radiocommunication.
    B. The master of the ship jointly with the station licensee.
    C. The third party for whom the message traffic was originated.
    D. The licensee of the ship station transmitting the messages.

133. Who determines when a ship station may transmit routine traffic destined for a coast or Government station in the maritime mobile service?

    A. The precedence of conventional radiocommunications is determined by FCC and international regulation.
    B. Ship stations must comply with instructions given by the coast or Government station.
    C. The order and time of transmission and permissible type of message traffic is decided by the licensed on-duty operator.
    D. Shipboard radio officers may transmit traffic when it will not interfere with on-going radiocommunications.

134. Under what circumstances may a ship or aircraft station interfere with a public coast station?

    A. In cases of distress.
    B. When it is necessary to transmit a message concerning the safety of navigation or important meteorological warnings.
    C. During periods of government-priority traffic handling.
    D. Under no circumstances during on-going radiocommunications.

135. What is the best way for a radio operator to minimize or prevent interference to other stations?

    A. Determine that a frequency is not in use by monitoring the frequency before transmitting.

    B. By changing frequency when notified that a radiocommunication causes interference.

    C. Reducing power to a level that will not affect other on-frequency communications.

    D. By using an omni-directional antenna pointed away from other stations.

136. What are the highest priority communications from ships at sea?

    A. Authorized government communications for which priority right has been claimed.

    B. Distress calls, and communications preceded by the international urgency and safety signals.

    C. Navigation and meteorological warnings.

    D. All critical message traffic authorized by the ship's master.

137. What is the international VHF digital selective calling channel?

    A. 500 kHz.

    B. 156.525 MHz.

    C. 156.35 MHz.

    D. 2182 kHz.

138. The primary purpose of bridge-to-bridge communications is:

    A. Navigational communications.

    B. Transmission of Captain's orders from the bridge.

    C. All short-range transmission aboard ship.

    D. Search and rescue emergency calls only.

139. The urgency signal has lower priority than:

    A. Security.

    B. Safety.

    C. Distress.

    D. Direction finding.

140. Under normal circumstances, what do you do if the transmitter aboard your ship is operating off-frequency, overmodulating, or distorting?

    A. Make a notation in station operating log.

    B. Reduce audio volume level.

    C. Stop transmitting.

    D. Reduce to low power.

141. What call should you transmit on Channel 16 if your ship is sinking?

    A. Urgency three times.

    B. Pan three times.

    C. Mayday three times.

    D. SOS three times.

142. Each cargo ship of the United States that is equipped with a radiotelephone station for compliance with Part II of Title III of the Communications Act shall while being navigated outside of a harbor or port keep a continuous and efficient watch on:

    A. Monitor all frequencies within the 2000 kHz to 27,500 kHz band used for communications.

    B. Both C and D.

    C. 156.8 MHz.

    D. 2182 kHz.

143. Radiotelephone stations required to keep logs of their transmissions must include:

    A. All of the below.

    B. Station callsigns with which communication took place.

    C. Name of operator on duty.

    D. Station, date, and time.

144. Cargo ships of 300 to 1600 gross tons should be able to transmit a minimum range of:

    A. 300 miles.

    B. 200 miles.

    C. 150 miles.

    D. 75 miles.

145. What is the priority of communications?

    A. Radio direction finding, distress, and safety.

    B. Distress, safety, radio direction finding, search, and rescue.

    C. Safety, distress, urgency, and radio direction finding.

    D. Distress, urgency, safety, and radio direction finding.

146. Which is a radiotelephony calling and distress frequency?

    A. 3113 kHz.

    B. 156.3 MHz.

    C. 2182 kHz.

    D. 500 kHz.

147. Portable ship radio transceivers operated as associated ship units:

    A. All of the below.

B. Must only communicate with the ship station with which it is associated or with associates portable ship units.

C. May not be used from shore without a separate license.

D. Must be operated on the safety and calling frequency 156.8 MHz (Channel 16) or a VHF intership frequency.

148. What is the most important practice that a radio operator must learn?

A. Always listen to 121.5 MHz.

B. Test a radiotelephone transmitter daily.

C. Operate with lowest power necessary.

D. Monitor the channel before transmitting.

149. What is required of a ship station that has established initial contact with another station on 2182 kHz or 156.800 MHz?

A. To expedite safety communications, the vessels must observe radio silence for two out of every 15 minutes.

B. Radiated power must be minimized so as not to interfere with other stations needing to use the channel.

C. The stations must change to an authorized working frequency for the transmission of messages.

D. The stations must check the radio channel for distress, urgency, and safety calls at least once every ten minutes.

150. On what frequency would a vessel normally call another ship station when using a radiotelephony emission?

A. On the vessel's unique working radio-channel assigned by the Federal Communications Commission.

B. On 2182 kHz or 156.800 MHz, unless the station knows the called vessel maintains a simultaneous watch on another intership working frequency.

C. On the appropriate calling channel of the ship station at 15 minutes past the hour.

D. Only on 2182 kHz in ITU (International Time Units) Region 2.

151. On what frequency should a ship station normally call a coast station when using a radiotelephony emission?

A. On 2182 kHz or 156.800 MHz at any time.

B. On any calling frequency internationally approved for use within ITU Region 2.

C. Calls should be initiated on the appropriate ship-to-shore working frequency of the coast station.

D. On a vacant radio channel determined by the licensed radio officer.

152. How is an associated vessel operating under the authority of another ship station license identified?

    A. Client vessels use the callsign of their parent plus the appropriate ITU regional indicator.

    B. By the callsign of the station with which it is connected and an appropriate unit designator.

    C. With any station callsign self-assigned by the operator of the associated vessel.

    D. All vessels are required to have a unique callsign sign issued by the Federal Communications Commission.

153. What is the power limitation of associated ship stations operating under the authority of a ship station license?

    A. Power is limited to 1 W.

    B. The minimum power necessary to complete the radiocommunications.

    C. Associated vessels are prohibited from operating under the authority granted to another station license.

    D. The power level authorized to the parent ship station.

154. Who has ultimate control of service at a ship's radio station?

    A. An appointed licensed radio operator who agrees to comply with all Radio Regulations in force.

    B. The Radio Officer-in-Charge authorized by the captain of the vessel.

    C. A holder of a First Class Radiotelegraph Certificate with a six months service endorsement.

    D. The master of the ship.

155. Under what circumstances may a coast station using telephony transmit a general call to a group of vessels?

    A. When identical traffic is destined for multiple mobile stations within range.

    B. When the vessels are located in international waters beyond 12 miles.

    C. When announcing or preceding the transmission of distress, urgency, safety, or other important messages.

    D. Under no circumstances.

156. How does a coast station notify a ship that it has a message for the ship?

    A. The coast station may transmit at intervals lists of callsigns in alphabetical order for which they have traffic.

    B. By establishing communications using the eight-digit maritime mobile service identification.

    C. The coast station changes to the vessel's known working frequency.

    D. By making a directed transmission on 2182 kHz or 156.800 MHz.

157. What is a bridge-to-bridge station?

    A. A VHF radio station located on a ship's navigational bridge or main control station that is used only for navigational communications.

    B. A portable ship station necessary to eliminate frequent application to operate a ship station on board different vessels.

C. An inland waterways and coastal radio station serving ship stations operating within the United States.

D. An internal communications system linking the wheel house with the ship's primary radio operating position and other integral ship control points.

158. What is the Automated Mutual-Assistance Vessel Rescue System?

A. A satellite-based distress and safety alerting program operated by the U.S. Coast Guard.

B. A coordinated radio direction finding effort between the Federal Communications Commission and U.S. Coast Guard to assist ships in distress.

C. An international system operated by the Coast Guard providing coordination of search and rescue efforts.

D. A voluntary organization of mariners who maintain radio watch on 500 kHz, 2182 kHz, and 156.800 MHz.

159. What are the radio watch requirements of a voluntary ship?

A. Radio watches must be maintained on the 156- to 158-MHz, 1600- to 4000-kHz, and 4000- to 23,000-kHz bands.

B. Radio watches are optional, but logs must be maintained of all medium, high-frequency, and VHF radio operation.

C. Radio watches must be maintained on 500 kHz, 2182 kHz, and 156.800 MHz, but no station logs are required.

D. Although licensees are not required to operate the ship radio station, general-purpose watches must be maintained if they do.

160. What is a Class "A" EPIRB?

A. A high efficiency audio amplifier.

B. An automatic, battery-operated emergency position-indicating radiobeacon that floats free of a sinking ship.

C. A satellite-based maritime distress- and safety-alerting system.

D. An alerting device that notifies mariners of imminent danger.

161. What is the minimum transmitter power level required by FCC for a medium-frequency transmitter aboard a compulsorily fitted vessel?

A. At least 25 W delivered into 50 Ω effective resistance when operated with a primary voltage of 13.6 Vdc.

B. The power predictably needed to communicate with the nearest public coast station operating on 2182 kHz.

C. At least 60 W PEP (Peak envelope power).

D. At least 100 W single side band suppressed carrier power.

162. How far from land may a small passenger vessel operate when equipped only with a VHF radiotelephone installation?

A. The vessel must remain within the communications range of the nearest coast station at all times.

B. No more than 20 nautical miles unless equipped with a reserve power supply.

C. No more than 100 nautical miles from the nearest land.

D. No more than 20 nautical miles from the nearest land if within the range of a VHF public coast or U.S. Coast Guard station.

163. How often must the radiotelephone installation aboard a small passenger boat be inspected?

A. A minimum of every 3 years, and when the ship is within 75 statute miles of an FCC field office.

B. At least once every five years.

C. When the vessel is first placed in service and every 2 years thereafter.

D. Equipment inspections are required at least once every 12 months.

164. What are the technical requirements of a VHF antenna system aboard a vessel?

A. The antenna must be constructed of corrosion-proof aluminum and capable of proper operation during an emergency.

B. The antenna must be capable of radiating a signal a minimum of 150 nautical miles on 156.8 MHz.

C. The antenna must be vertically polarized and nondirectional.

D. The antenna must provide an amplification factor of at least 2.1 dbi.

165. In the International Phonetic Alphabet, the letters D, N, and O are represented by the words:

A. Delta, Neptune, Olive.

B. December, Nebraska, Olive.

C. Denmark, Neptune, Oscar.

D. Delta, November, Oscar.

166. What is selective calling?

A. A telegraphy transmission directed only to another specific radiotelegraph station.

B. An electronic device that uses a discriminator circuit to filter out unwanted signals.

C. A radiotelephony communication directed at a particular ship station.

D. A coded transmission directed to a particular ship station.

167. What authorization is required to operate a 350-W PEP maritime voice station on frequencies below 30 MHz aboard a small noncommercial pleasure vessel?

A. Marine Radio Operator Permit.

B. Restricted Radiotelephone Operator Permit.

C. General Radiotelephone Operator License.

D. Third Class Radiotelegraph Operator's Certificate.

168. What is the proper procedure for making a correction in the station log?

A. Rewrite the new entry in its entirety directly below the incorrect notation and initial the change.

B. The original person making the entry must strike out the error, initial the correction and indicate the date of correction.

C. The mistake may be erased and the correction made and initialed only by the radio operator making the original error.

D. The ship's master must be notified, approve, and initial all changes to the station log.

169. What is the purpose of the international radiotelephone alarm signal?

A. To actuate automatic devices giving an aural alarm to attract the attention of the operator where there is no listening watch on the distress frequency.

B. To alert radio officers monitoring watch frequencies of a forthcoming distress, urgency of safety message.

C. To call attention to the upcoming transmission of an important meteorological warning.

D. To notify nearby ships of the loss of a person or persons overboard.

170. On what frequencies does the Communications Act require radio watches by compulsory radiotelephone stations?

A. Watches are required on 2182 kHz and 156.800 MHz.

B. On all frequencies between 405 to 535 kHz, 1605 to 3500 kHz and 156 to 162 MHz.

C. Continuous watch is required on 2182 kHz only.

D. Watches are required on 500 kHz and 2182 kHz.

171. What FCC authorization is required to operate a VHF transmitter on board a vessel voluntarily equipped with radio and sailing on a domestic voyage?

A. General Radiotelephone Operator License.

B. Restricted Radiotelephone Operator Permit.

C. Marine Radio Operator Permit.

D. No radio operator license or permit is required.

172. What is the minimum radio operator requirement for ships subject to the Great Lakes Radio Agreement?

A. Restricted Radiotelephone Operator Permit.

B. Marine Radio Operator Permit.

C. General Radiotelephone Operator License.

D. Third Class Radiotelegraph Operator's Certificate.

173. What is the proper procedure for testing a radiotelephone installation?

A. Short tests must be confined to a single working frequency and must never be conducted.

    B. Permission for the voice test must be requested and received from the nearest public coast station.

    C. A dummy antenna must be used to ensure that the test will not interfere with ongoing communications.

    D. Transmit the station's callsign, followed by the word "test" on the radio channel being used for the test.

174. What should a station operator do before making a transmission?

    A. Ask if the frequency is in use.

    B. Check transmitting equipment to be certain it is properly calibrated.

    C. Except for the transmission of distress calls, determine that the frequency is not in use by monitoring the frequency before transmitting.

    D. Transmit a general notification that the operator wishes to utilize the channel.

175. Where must the principal radiotelephone operating position be installed in a ship station?

    A. At the level of the main wheel house or at least one deck above the ship's main deck.

    B. In the chart room, master's quarters or wheel house.

    C. In the room or an adjoining room from which the ship is normally steered while at sea.

    D. At the principal radio operating position of the vessel.

176. What is the antenna requirement of a radiotelephone installation aboard a passenger vessel?

    A. All antennas must be tested and the operational results logged at least once during each voyage.

    B. The antenna must be vertically polarized and as nondirectional and efficient as is practicable for the transmission and reception of ground waves over seawater.

    C. An emergency reserve antenna system must be provided for communications on 156.8 MHz.

    D. The antenna must be located a minimum of 15 meters from the radiotelegraph antenna.

177. Where must ship station logs be kept during a voyage?

    A. All logs are turned over to the ship's master when the radio operator goes off duty.

    B. In the personal custody of the licensed commercial radio operator.

    C. They must be secured in the vessel's strong-box for safekeeping.

    D. At the principal radiotelephone-operating position.

178. How long should station logs be retained when there are entries relating to distress or disaster situations?

A. For a period of one year from the date of entry.

B. For a period of three years from the date of entry unless notified by the FCC.

C. Indefinitely, or until destruction is specifically authorized by the U.S. Coast Guard.

D. Until authorized by the Commission in writing to destroy them.

179. Who is responsible for the proper maintenance of station logs?

A. The ship's master and the station licensee.

B. The commercially licensed radio operator in charge of the station.

C. The station licensee.

D. The station licensee and the radio operator in charge of the station.

180. Who may be granted a ship station license in the maritime service?

A. The owner or operator of a vessel, or their subsidiaries.

B. Vessels that have been inspected and approved by the U.S. Coast Guard and Federal Communications Commission.

C. Only FCC licensed operators holding a First or Second Class Radiotelegraph Operator's Certificate or the General Radiotelephone Operator License.

D. Anyone, including foreign governments.

181. What is a distress communication?

A. An official radiocommunications notification of approaching navigational or meteorological hazards.

B. Radiocommunications, which, if delayed, will adversely affect the safety of life or property.

C. Communications indicating that the calling station has a very urgent message concerning safety.

D. An internationally recognized communication indicating that the sender is threatened by grave and imminent danger and requests immediate assistance.

182. What is the Communication Act's definition of a "passenger ship?"

A. A vessel of any nation that has been inspected and approved as a passenger-carrying vessel.

B. Any ship transporting more than six passengers for hire.

C. A vessel that carries or is licensed or certificated to carry more than 12 passengers.

D. Any ship that is used primarily in commerce for transporting persons to and from harbors or ports.

183. By international agreement, which ships must carry radio equipment for the safety of life at sea?

    A. All cargo ships of more than 100 gross tons.

    B. Cargo ships of more than 100 gross tons and passenger vessels on international deep-sea voyages.

    C. All ships that travel more than 100 miles out to sea.

    D. Cargo ships of more than 300 gross tons and vessels carrying more than 12 passengers.

184. When using an SSB station on 2182 kHz or VHF-FM on Channel 16:

    A. All of the below.

    B. Once contact is established, you must switch to a working frequency.

    C. If contact is not made, you must wait at least 2 minutes before repeating the call.

    D. Preliminary call must not exceed 30 seconds.

185. A ship station using VHF bridge-to-bridge Channel 13:

    A. Does not need to identify itself within 100 miles from shore.

    B. May be identified by the name of the ship in lieu of callsign.

    C. Must be identified by callsign and name of vessel.

    D. May be identified by callsign and country of origin.

186. What is the average range of VHF marine transmissions?

    A. 10 miles.

    B. 20 miles.

    C. 50 miles.

    D. 150 miles.

187. How should the 2182 kHz auto-alarm be tested?

    A. Only under U.S. Coast Guard authorization.

    B. On 2182 kHz into antenna.

    C. On a different frequency into dummy load.

    D. On a different frequency into antenna.

188. International laws and regulations require a silent period on 2182 kHz:

    A. Both C and D below.

    B. For the first minute of every quarter-hour.

    C. For three minutes immediately after the half-hour.

    D. For three minutes immediately after the hour.

189. If a ship sinks, what device is designed to float free of the mother ship, is turned on automatically and transmits a distress signal?

    A. Auto alarm keyer on any frequency.

    B. Bridge-to-bridge transmitter on 2182 kHz.

    C. EPIRB on 2182 kHz and 405.025 kHz.

    D. EPIRB on 121.5 MHz/243 MHz or 406.025 MHz.

190. The radiotelephone distress message consists of:

    A. All of the below.
    B. Nature of distress and kind of assistance desired.
    C. Particulars of its position, latitude and longitude, and other information which might facilitate rescue, such as length, color and type of vessel, number of persons on board.
    D. Mayday spoken three times, the callsign, and the name of vessel in distress.

191. When are EPIRB batteries changed?

    A. Whenever voltage drops to less than 50% of full charge.
    B. After emergency use; every 12 months when not used.
    C. After emergency use; as per manufacturers instructions marked on outside of transmitter with month and year replacement date.
    D. After emergency use; after battery life expires.

192. When may a bridge-to-bridge transmission be more than 1 W?

    A. Both C and D.
    B. When calling the Coast Guard.
    C. When rounding a bend in a river or traveling in a blind spot.
    D. When broadcasting a distress message.

193. What must be in operation when no operator is standing watch on a compulsory radio equipped vessel while out at sea?

    A. Radiotelegraph transceiver set to 2182 kHz.
    B. Distress-Alert signal device.
    C. Indicating Radio Beacon signals.
    D. An auto alarm.

194. When is it legal to transmit high power on Channel 13?

    A. All of the below.
    B. During an emergency.
    C. In a blind situation, such as rounding a bend in a river.
    D. Failure of vessel being called to respond.

195. What transmitting equipment is authorized for use by a station in the maritime services?

    A. Transceivers and transmitters that meet all ITU specifications for use in maritime mobile service.
    B. Equipment that has been inspected and approved by the U.S. Coast Guard.
    C. Unless specifically excepted, only transmitters that are type accepted by the Federal Communications Commission for Part 80 operations.
    D. Transmitters that have been certified by the manufacturer for maritime use.

196. Which commercial radio operator license is required to install a VHF transmitter in a voluntarily equipped ship station?
   A. A General Radiotelephone Operator License.
   B. A Restricted Radiotelephone Operator Permit or higher class of license.
   C. None, if installed by, or under the supervision of, the licensee of the ship station and no modifications are made to any circuits.
   D. A Marine Radio Operator Permit or higher class of license.

197. Which commercial radio operator license is required to operate a fixed tuned ship radar station with external controls?
   A. No radio operator authorization is required.
   B. Either a First or Second Class Radiotelegraph certificate or a General Radiotelephone Operator License.
   C. A Marine Radio Operator Permit or higher.
   D. A radio operator certificate containing a Ship Radar Endorsement.

198. What are the radio operator requirements of a small passenger ship carrying more than six passengers equipped with a 1000-W carrier power radiotelephone station?
   A. The operator must hold a GMDSS Radio Operator's License.
   B. The operator must hold a Restricted Radiotelephone Operator Permit or higher class license.
   C. The operator must hold a Marine Radio Operator Permit or higher class license.
   D. The operator must hold a General Radiotelephone Operator or higher class license.

199. What are the radio operator requirements of a cargo ship with a 1000-W peak-envelope-power radiotelephone station?
   A. The operator must hold a GMDSS Radio Maintainer's License.
   B. The operator must hold a Marine Radio Operator Permit or higher class license.
   C. The operator must hold a Restricted Radiotelephone Operator Permit or higher class license.
   D. The operator must hold a General Radiotelephone Operator License or higher class license.

200. When may the operator of a ship radio station allow an unlicensed person to speak over the transmitter?
   A. During the hours that the radio office is normally off duty.
   B. When under the supervision of the licensed operator.
   C. When the station power does not exceed 200 W peak envelope power.
   D. At no time. Only commercially licensed radio operators may modulate the transmitting apparatus.

201. What are the service requirements of all ship stations?

A. Reserve antennas, emergency power sources and alternate communications installations must be available.
B. All ship stations must maintain watch on 500 kHz, 2182 kHz, and 156.800 MHz.
C. Public correspondence services must be offered for any person during the hours the radio operator is normally on duty.
D. Each ship station must receive and acknowledge all communications with any station in the maritime mobile service.

202. What type of communications may be exchanged by radioprinter between authorized private coast stations and ships of less than 1600 gross tons?

A. There are no restrictions.
B. Only those communications which concern the business and operational needs of vessels.
C. All communications providing they do not exceed 3 minutes after the stations have established contact.
D. Public correspondence service may be provided on voyages of more than 24 hours.

203. How do the rules define "navigational communications"?

A. Radio signals consisting of weather, sea conditions, notices to mariners and potential dangers.
B. Telecommunications pertaining to the guidance of maritime vessels in hazardous waters.
C. Important communications concerning the routing of vessels during periods of meteorological crisis.
D. Safety communications pertaining to the maneuvering or directing of vessels movements.

204. How do the FCC's Rules define a power-driven vessel?

A. A vessel moved by mechanical equipment at a rate of 5 knots or more.
B. A watercraft containing a motor with a power rating of at least 3 HP.
C. Any ship propelled by machinery.
D. A ship that is not manually propelled or under sail.

205. What is a "passenger carrying vessel" when used in reference to the Great Lakes Radio Agreement?

A. A ship that is used primarily for transporting persons and goods to and from domestic harbors or ports.
B. Any ship, the principal purpose of which is to ferry persons on the Great Lakes and other inland waterways.
C. Any ship carrying more than six passengers for hire.
D. A vessel that is licensed or certificated to carry more than 12 passengers.

206. Which of the following statements is true as to ships subject to the Safety Convention?

    A. Cargo ships are FCC inspected on an annual basis while passenger ships undergo U.S. Coast Guard inspections every six months.

    B. A cargo ship is any ship that is not licensed or certificated to carry more than 12 passengers.

    C. Passenger ships carry six or more passengers for hire, as opposed to transporting merchandise.

    D. A cargo ship participates in international commerce by transporting goods between harbors.

207. What is the order of priority of radiotelephone communications in the maritime services?

    A. Government precedence, messages concerning safety of life and protection of property and traffic concerning grave and imminent danger.

    B. Navigation hazards, meteorological warning, priority traffic.

    C. Alarm, radio-direction finding, and health and welfare communications.

    D. Distress calls and signals, followed by communications that are preceded by urgency and safety signals.

208. When may a shipboard radio operator make a transmission in the maritime services not addressed to a particular station or stations?

    A. When the radio officer is more than 12 miles from shore.

    B. Only when specifically authorized by the master of the ship.

    C. Only during the transmission of distress, urgency or safety signals or messages, or to test the station.

    D. General CQ calls may only be made when the operator is off duty and another operator is on watch.

209. What action must be taken by the owner or operator of a vessel who changes its name?

    A. Written confirmation must be obtained from the U.S. Coast Guard.

    B. The Federal Communications Commission in Gettysburg, PA, must be notified in writing.

    C. The Engineer-in-Charge of the nearest FCC field office must be informed.

    D. A Request for Ship License Modification (RSLM) must be submitted to the FCC's licensing facility.

210. What traffic management service is operated by the U.S. Coast Guard in certain designated water areas to prevent ship collisions, groundings and environmental harm?

    A. Interdepartmental harbor and port patrol (IHPP).

    B. Ship movement and safety agency (SMSA).

    C. Vessel traffic service (VTS).

    D. Water safety management bureau (WSMB).

Answer sheet

1. B	2. A	3. D	4. D	5. B
6. C	7. C	8. D	9. A	10. C
11. A	12. D	13. B	14. B	15. C
16. B	17. D	18. B	19. B	20. C
21. C	22. D	23. A	24. D	25. A
26. D	27. C	28. B	29. A	30. D
31. A	32. D	33. C	34. C	35. B
36. A	37. A	37. A	39. D	40. A
41. C	42. C	43. D	44. B	45. D
46. B	47. C	48. B	49. D	50. C
51. C	52. C	53. C	54. D	55. A
56. B	57. A	58. A	59. B	60. A
61. C	62. D	63. D	64. D	65. C
66. A	67. A	68. A	69. A	70. B
76. B	77. B	78. D	79. B	80. A
71. A	72. D	73. C	74. B	75. B
81. A	82. D	83. A	84. D	85. B
86. A	87. D	88. C	89. D	90. A
91. A	92. B	93. B	94. C	95. B
96. C	97. D	98. B	99. D	100. B
101. B	102. C	103. A	104. A	105. C
106. D	107. A	108. D	109. D	110. B
111. D	112. B	113. C	114. A	115. C
116. B	117. D	118. B	119. B	120. A
121. C	122. A	123. A	124. B	125. A
126. C	127. A	128. D	129. C	130. B
131. B	132. D	133. B	134. A	135. A
136. B	137. B	138. A	139. C	140. C
141. C	142. B	143. A	144. C	145. D
146. C	147. A	148. D	149. C	150. B
151. C	152. B	153. A	154. D	155. C
156. A	157. A	158. C	159. D	160. B
161. C	162. D	163. B	164. C	165. D
166. D	167. B	168. B	169. A	170. A
171. D	172. B	173. D	174. C	175. C
176. B	177. D	178. B	179. D	180. A
181. D	182. C	183. D	184. A	185. B
186. B	187. C	188. A	189. D	190. A
191. C	192. A	193. D	194. A	195. C
196. C	197. A	198. A	199. B	200. B
201. D	202. B	203. D	204. C	205. C
206. B	207. D	208. C	209. B	210. C

6

Programmed review
Questions related to
transmitters and receivers

Transmitters

Start with Block 1. Pick the answer you believe is correct. Go to the next block and check your answer. All answers are in italics. There is only one choice for each block.

Block 1

Here is your first question: The band of frequencies least susceptible to atmospheric noise and interference is:

A. 300 to 3000 MHz.
B. 3 to 30 MHz.
C. 300 to 3000 kHz.
D. 30 to 300 kHz.

Block 2

The correct answer is A.

Here is your next question: What is the best time of day for transequatorial propagation?

A. Transequatorial propagation only works at night.
B. Afternoon or early evening.
C. Noon.
D. Morning.

Block 3

The correct answer is B.

Here is your next question: What is transequatorial propagation?

A. Propagation between any two stations at the same latitude.
B. Propagation between two continents by way of ducts along the magnetic equator.
C. Propagation between two points on the magnetic equator.
D. Propagation between two points at approximately the same distance north and south of the magnetic equator.

Block 4

The correct answer is D.

Here is your next question: What ferrite device can be used instead of a duplexer to isolate a microwave transmitter and receiver when both are connected to the same antenna?

A. Simplex.
B. Magnetron.
C. Circulator.
D. Isolator.

Block 5

The correct answer is C.

Here is your next question: In facsimile, how are variations in picture brightness and darkness converted into voltage variations?

A. With an optoisolator.
B. With a photodetector.
C. With a Hall-effect transistor.
D. With an LED.

Block 6

The correct answer is B.

Here is your next question: What is the approximate transmission time for a facsimile picture transmitted by a radio station?

A. $\frac{1}{60}$ second per frame at 240 lines per minute.
B. Approximately 6 seconds per frame at 240 lines per minute.
C. Approximately 3.3 minutes per frame at 240 lines per minute.
D. Approximately 6 minutes per frame at 240 lines per minute.

Block 7

The correct answer is C.

Here is your next question: What is facsimile?

A. The transmission of printed pictures for permanent display on paper.
B. The transmission of video by television.
C. The transmission of still pictures by slow-scan television.
D. The transmission of characters by radioteletype that form a picture when printed.

Block 8

The correct answer is A.

Here is your next question: VHF ship station transmitters must have the capability of reducing carrier power to:

A. 50 W.
B. 25 W.
C. 10 W.
D. 1 W.

Block 9

The correct answer is D.

Here is your next question: What is the antenna requirement of a radiotelephone installation aboard a passenger vessel?

A. All antennas must be tested and the operational results logged at least once during each voyage.
B. The antenna must be vertically polarized and as nondirectional and efficient as is practicable for the transmission and reception of ground waves over seawater.
C. An emergency reserve antenna system must be provided for communications on 156.8 MHz.
D. The antenna must be located a minimum of 15 meters from the radiotelegraph antenna.

Block 10

The correct answer is B.

Here is your next question: Who has ultimate control of service at a ship's radio station?

A. An appointed licensed radio operator who agrees to comply with all radio regulations in force.
B. The Radio Officer-in-Charge authorized by the captain of the vessel.
C. A holder of a First Class Radiotelegraph Certificate with a six months service endorsement.
D. The master of the ship.

Block 11

The correct answer is D.

Here is your next question: The average range for VHF communications is:

A. 100 miles.
B. 30 miles.
C. 15 miles.
D. 5 miles.

Block 12

The correct answer is B.

Here is your next question: What is the maximum range for signals using transequatorial propagation?

A. About 7500 miles.
B. About 5000 miles.
C. About 2500 miles.
D. About 1000 miles.

Block 13

The correct answer is B.

Here is your next question: How does the bandwidth of the transmitted signal affect selective fading?

A. The receiver bandwidth determines the selective fading effect.
B. It is equally pronounced at both narrow and wide bandwidths.
C. It is more pronounced at narrow bandwidths.
D. It is more pronounced at wide bandwidths.

Block 14

The correct answer is D.

Here is your next question: A 25-MHz amplitude-modulated transmitter's actual carrier frequency is 25.00025 MHz without modulation and is 24.99950 MHz when modulated. What statement is true?

A. If the authorized frequency tolerance is 0.005% for the 25-MHz band this transmitter is operating legally.
B. Modulation should not change carrier frequency.
C. If the allowed frequency tolerance is 0.002%, this is an illegal transmission.
D. If the allowed frequency tolerance is 0.001%, this is an illegal transmission.

Block 15

The correct answer is A.

Here is your next question: What is the term for the transmission of printed pictures by radio for the purpose of a permanent display?

A. ACSSB.
B. Xerography.
C. Facsimile.
D. Television.

Block 16

The correct answer is C.

Here is your next question: What is the modern standard scan rate for a facsimile picture transmitted by a radio station?

A. The modern standard is 60 lines per second.

B. The modern standard is 150 lines per second.
C. The modern standard is 50 lines per minute.
D. The modern standard is 240 lines per minute.

Block 17

The correct answer is D.

Here is your next question: If your transmitter is producing spurious harmonics or is operating at a deviation from the technical requirements of the station authorization:

A. reduce power immediately.
B. cease transmission.
C. repair problem within 24 hours.
D. continue operating until returning to port.

Block 18

The correct answer is B.

Here is your next question: What is the average range of VHF marine transmissions?

A. 10 miles.
B. 20 miles.
C. 50 miles.
D. 150 miles.

Block 19

The correct answer is B.

Here is your next question: What transmitting equipment is authorized for use by a station in the maritime services?

A. Transceivers and transmitters that meet all ITU specifications for use in maritime mobile service.
B. Equipment that has been inspected and approved by the U.S. Coast Guard.
C. Unless specifically excepted, only transmitter types that are accepted by the Federal Communications Commission for Part 80 operations.
D. Transmitters that haven't been certified by the manufacturer for maritime use.

Block 20

The correct answer is C.

Here is your next question: What is a maritime mobile repeater station?

A. A one-way low-power communications system used in the maneuvering of vessels.
B. A mobile radio station which links two or more public coast stations.
C. An automatic on-board radio station which facilitates the transmissions of safety communications aboard ship.
D. A fixed land station used to extend the communications range of ship and coast stations.

Block 21

The correct answer is D.

Here is your next question: What is the effective radiated power of a repeater with 200-W transmitter power output, 4-dB feedline loss, 4-dB duplexer and circulator loss, and 10-dB antenna gain?

A. 260 W, assuming the antenna gain is referenced to a half-wave dipole.
B. 126 W, assuming the antenna gain is referenced to a half-wave dipole.
C. 2000 W, assuming the antenna gain is referenced to a half-wave dipole.
D. 317 W, assuming the antenna gain is referenced to a half-wave dipole.

Block 22

The correct answer is D.

Here is your next question: What is the effective radiated power of a repeater with 120-W transmitter power output, 5-dB feedline loss, 4-dB duplexer and circulator loss, and 6-dB antenna gain?

A. 379 watts, assuming the antenna gain is referenced to a half-wave dipole.
B. 60 W, assuming the antenna gain is referenced to a half-wave dipole.
C. 240 W, assuming the antenna gain is referenced to a half-wave dipole.
D. 601 W, assuming the antenna gain is referenced to a half-wave dipole.

Block 23

The correct answer is B.

Here is your next question: What is the effective radiated power of a repeater with 100-W transmitter power output, 4-dB feedline loss, 3-dB duplexer and circulator loss, and 7-dB antenna gain?

A. 100 W, assuming the antenna gain is referenced to a half-wave dipole.
B. 25 W, assuming the antenna gain is referenced to a half-wave dipole.
C. 400 W, assuming the antenna gain is referenced to a half-wave dipole.
D. 631 W, assuming the antenna gain is referenced to a half-wave dipole.

Block 24

The correct answer is A.

Here is your next question: What is the effective radiated power of a repeater with 75-W transmitter power output, 4-dB feedline loss, 3-dB duplexer and circulator loss, and 10-dB antenna gain?

A. 150 W, assuming the antenna gain is referenced to a half-wave dipole.
B. 18.75 W, assuming the antenna gain is referenced to a half-wave dipole.
C. 75 W, assuming the antenna gain is referenced to a half-wave dipole.
D. 600 W, assuming the antenna gain is referenced to a half-wave dipole.

Block 25

The correct answer is A.

Here is your next question: What is the effective radiated power of a repeater with 50-W transmitter power output, 4-dB feedline loss, 3-dB duplexer and circulator loss, and 6-dB antenna gain?

A. 69.9 W, assuming the antenna gain is referenced to a half-wave dipole.
B. 251 W, assuming the antenna gain is referenced to a half-wave dipole.
C. 39.7 W, assuming the antenna gain is referenced to a half-wave dipole.
D. 158 W, assuming the antenna gain is referenced to a half-wave dipole.

Block 26

The correct answer is C.

Here is your next question: What is the term used to refer to the condition where the signals from a very strong station are superimposed on other signals being received?

A. Capture effect.
B. Receiver quieting.
C. Cross-modulation interference.
D. Intermodulation distortion.

Block 27

The correct answer is C.

Here is your next question: How can intermodulation interference between two transmitters in close proximity often be reduced or eliminated?

A. By installing a low-pass filter in the antenna feedline.
B. By installing a band-pass filter in the antenna feedline.
C. By installing a terminated circulator or ferrite isolator in the feedline to the transmitter and duplexer.
D. By using a class-C final amplifier with high driving power.

Block 28

The correct answer is C.

Here is your next question: What is the name of the condition that occurs when the signals of two transmitters in close proximity mix together in one or both of their final amplifiers, and unwanted signals at the sum and difference frequencies of the original transmissions are generated?

A. Intermodulation interference.
B. Adjacent-channel interference.
C. Neutralization.
D. Amplifier desensitization.

Block 29

The correct answer is A.

Here is your next question: What is the most the actual transmitter frequency could differ from a reading of 462,100,000 Hz on a frequency counter with a time base accuracy of ±0.1 ppm?

A. 0.2 MHz.
B. 462.1 Hz.
C. 0.1 MHz.
D. 46.21 Hz.

Block 30

The correct answer is D.

Here is your next question: What is the most the actual transmitter frequency could differ from a reading of 156,520,000 Hz on a frequency counter with a time base accuracy of ±10 ppm?

A. 1565.20 Hz.
B. 156.52 kHz.
C. 10 Hz.
D. 146.42 Hz.

Block 31

The correct answer is A.

Here is your next question: What is the effective radiated power of a repeater with 150-W transmitter power output, 4-dB feedline loss, 3-dB duplexer and circulator loss, and 7-dB antenna gain?

A. 150 W, assuming the antenna gain is referenced to a half-wave dipole.
B. 600 W, assuming the antenna gain is referenced to a half-wave dipole.
C. 37.5 W, assuming the antenna gain is referenced to a half-wave dipole.
D. 946 W, assuming the antenna gain is referenced to a half-wave dipole.

Block 32

The correct answer is A.

Here is your next question: What is the effective radiated power of a repeater with 100-W transmitter power output, 5-dB feedline loss, 4-dB duplexer and circulator loss, and 10-dB antenna gain?

A. 1260 W, assuming the antenna gain is referenced to a half-wave dipole.
B. 12.5 W, assuming the antenna gain is referenced to a half-wave dipole.
C. 126 W, assuming the antenna gain is referenced to a half-wave dipole.
D. 800 W, assuming the antenna gain is referenced to a half-wave dipole.

Block 33

The correct answer is C.

Here is your next question: What is the effective radiated power of a repeater with 75-W transmitter power output, 5-dB feedline loss, 4-dB duplexer and circulator loss, and 6-dB antenna gain?

A. 23.7 W, assuming the antenna gain is referenced to a half-wave dipole.
B. 150 W, assuming the antenna gain is referenced to a half-wave dipole.

C. 237 W, assuming the antenna gain is referenced to a half-wave dipole.
D. 37.6 W, assuming the antenna gain is referenced to a half-wave dipole.

Block 34

The correct answer is D.

Here is your next question: What is the effective radiated power of a repeater with 50-W transmitter power output, 5-dB feedline loss, 4-dB duplexer and circulator loss, and 7-dB antenna gain?

A. 69.9 W, assuming the antenna gain is referenced to a half-wave dipole.
B. 31.5 W, assuming the antenna gain is referenced to a half-wave dipole.
C. 315 W, assuming the antenna gain is referenced to a half-wave dipole.
D. 300 W, assuming the antenna gain is referenced to a half-wave dipole.

Block 35

The correct answer is B.

Here is your next question: If there are too many harmonics from a transmitter, check the:

A. Any of the above.
B. shielding.
C. tuning of circuits.
D. coupling.

Block 36

The correct answer is A.

Here is your next question: How can even-order harmonics be reduced or prevented in transmitter amplifier design?

A. By operating class AB.
B. By operating class C.
C. By using a push-pull amplifier.
D. By using a push-push amplifier.

Block 37

The correct answer is C.

Here is your next question: How does intermodulation interference between two transmitters usually occur?

A. When the signals from the transmitters are reflected in phase from the airplanes passing overhead.
B. When they are in close proximity and the signals cause feedback in one or both of their final amplifiers.
C. When they are in close proximity and the signals mix in one or both of their final amplifiers.

D. When the signals from the transmitters are reflected out of phase from airplanes passing overhead.

Block 38

The correct answer is C.

Here is your next question: What is the most the actual transmitter frequency could differ from a reading of 462,100,000 Hz on a frequency counter with a time base accuracy of ±10 ppm?

A. 462.1 Hz.
B. 4621 Hz.
C. 10 Hz.
D. 10 MHz.

Block 39

The correct answer is B.

Here is your next question: What is the most the actual transmitter frequency could differ from a reading of 462,100,000 Hz on a frequency counter with a time base accuracy of ±1.0 ppm?

A. 462.1 Hz.
B. 1.0 MHz.
C. 10 Hz.
D. 46.21 MHz.

Block 40

The correct answer is A.

Here is your next question: What is the most the actual transmitter frequency could differ from a reading of 156,520,000 Hz on a frequency counter with a time base accuracy of ±0.1 ppm?

A. 1.5652 kHz.
B. 1.4652 Hz.
C. 0.1 MHz.
D. 15.652 Hz.

Block 41

The correct answer is D.

Here is your next question: What term describes a wide-bandwidth communications systems in which the RF carrier varies according to some predetermined sequence?

A. Spread-spectrum communication.
B. Time-domain frequency modulation.
C. SITOR.
D. Amplitude-compandored single sideband.

Block 42

The correct answer is A.

Here is your next question: What is the approximate frequency of the pilot tone in an amplitude-compandored single-sideband system?

A. 3 kHz.
B. 455 kHz.
C. 5 MHz.
D. 1 kHz.

Block 43

The correct answer is A.

Here is your next question: What is meant by compandoring?

A. Detecting and demodulating a single-sideband signal by converting it to a pulse-modulated signal.
B. Combining amplitude and frequency modulation to produce a single-sideband signal.
C. Using an audio-frequency signal to produce pulse-length modulation.
D. Compressing speech at the transmitter and expanding it at the receiver.

Block 44

The correct answer is D.

Here is your next question: What is one way that voice is transmitted in a pulse-width modulation system?

A. The number of standard pulses per second varies depending on the voice waveform at that instant.
B. A standard pulse is varied in duration by an amount depending on the voice waveform at that instant.
C. The position of a standard pulse is varied by an amount depending on the voice waveform at that instant.
D. A standard pulse is varied in amplitude by an amount depending on the voice waveform at that instant.

Block 45

The correct answer is B.

Here is your next question: In a pulse-position modulation system, what parameter does the modulating signal vary?

A. The time at which each pulse occurs.
B. The duration of the pulses.
C. Both the frequency and amplitude of the pulses.
D. The number of pulses per second.

Block 46

The correct answer is A.

Here is your next question: What is the approximate dc input power to a class-C RF power amplifier stage in an RTTY transmitter when the PEP output power is 1000 W?

A. Approximately 2000 W.
B. Approximately 1667 W.
C. Approximately 1250 W.
D. Approximately 850 W.

Block 47

The correct answer is C.

Here is your next question: In a single-sideband phone signal, what determines the PEP-to-average power ratio?

A. The amplifier power.
B. The speech characteristics.
C. The degree of carrier suppression.
D. The frequency of the modulating signal.

Block 48

The correct answer is B.

Here is your next question: How can a single-sideband phone signal be produced?

A. By producing a double-sideband signal with a balanced modulator and then removing the unwanted sideband by neutralization.
B. By producing a double-sideband signal with a balanced modulator and then removing the unwanted sideband by mixing.
C. By producing a double-sideband signal with a balanced modulator and then removing the unwanted sideband by heterodyning.
D. By producing a double-sideband signal with a balanced modulator and then removing the unwanted sideband by filtering.

Block 49

The correct answer is D.

Here is your next question: How can an FM-phone signal be produced?

A. By using a balanced modulator on an oscillator.
B. By using a reactance modulator on an oscillator.
C. By modulating the supply voltage to a class-C amplifier.
D. By modulating the supply voltage to a class-B amplifier.

Block 50

The correct answer is B.

Here is your next question: What is emission F3F?

A. Television.
B. RTTY.
C. Facsimile.
D. Modulated CW.

Block 51

The correct answer is A.

Here is your next question: What is emission A3F?

A. Modulated CW.
B. SSB.
C. Television.
D. RTTY.

Block 52

The correct answer is C.

Here is your next question: What is emission F3F?

A. Facsimile.
B. RTTY.
C. Slow-Scan TV.
D. Voice transmission.

Block 53

The correct answer is A.

Here is your next question: What type of emission is produced when an amplitude modulated transmitter is modulated by a facsimile signal?

A. F3C.
B. F3F.
C. A3C.
D. A3F.

Block 54

The correct answer is C.

Here is your next question: What is the term for the time required for the current in an RL circuit to build up to 63.2% of the maximum value?

A. One exponential rate.
B. A time factor of one.
C. An exponential period of one.
D. One time constant.

Block 55

The correct answer is D.

Here is your next question: What is the meaning of the term *time constant* of an RL circuit?

A. The time required for the voltage in the circuit to build up to 36.8% of the maximum value.
B. The time required for the current in the circuit to build up to 63.2% of the maximum value.
C. The time required for the voltage in the circuit to build up to 63.2% of the maximum value.
D. The time required for the current in the circuit to build up to 36.8% of the maximum value.

Block 56

The correct answer is B.

Here is your next question: What does the photoconductive effect in crystalline solids produce a noticeable change in?

A. The resistance of the solid.
B. The specific gravity of the solid.
C. The inductance of the solid.
D. The capacitance of the solid.

Block 57

The correct answer is A.

Here is your next question: What is an optoisolator?

A. An LED and a solar cell.
B. An LED and a capacitor.
C. A P-N junction that develops an excess positive charge when exposed to light.
D. An LED and a phototransistor.

Block 58

The correct answer is D.

Here is your next question: How many voice transmissions can be packed into a given frequency band for amplitude-compandored single-sideband systems over conventional FM-phone systems?

A. 16.
B. 8.
C. 4.
D. 2.

Block 59

The correct answer is C.

Here is your next question: What is the purpose of a pilot tone in an amplitude-compandored single-sideband system?

A. It acts as a beacon to indicate the present propagation characteristic of the band.
B. It permits rapid change of frequency to escape high-powered interference.
C. It replaces the suppressed carrier at the receiver.
D. It permits rapid tuning of a mobile receiver.

Block 60

The correct answer is D.

Here is your next question: What is amplitude-compandored single sideband?

A. Single sideband incorporating speech expansion at the transmitter and speech compression at the receiver.
B. Single sideband incorporating speech compression at the transmitter and speech expansion at the receiver.
C. Reception of single sideband with a conventional FM receiver.
D. Reception of single sideband with a conventional CW receiver.

Block 61

The correct answer is B.

Here is your next question: Why is the transmitter peak power in a pulse modulation system much greater than its average power?

A. The signal reaches peak amplitude only when the pulses are also amplitude modulated.
B. The signal reaches peak amplitude only when voltage spikes are generated within the modulator.
C. The signal reaches peak amplitude only when voice modulated.
D. The signal duty cycle is less than 100%.

Block 62

The correct answer is D.

Here is your next question: What is the type of modulation in which the modulating signal varies the duration of the transmitted pulse?

A. Pulse-height modulation.
B. Pulse-width modulation.
C. Frequency modulation.
D. Amplitude modulation.

Block 63

The correct answer is B.

Here is your next question: What is the approximate dc input power to a class-B RF power amplifier stage in an FM-phone transmitter when the PEP output is 1500 W?

A. Approximately 3000 W.
B. Approximately 2500 W.

C. Approximately 1765 W.
D. Approximately 900 W.

Block 64

The correct answer is B.

Here is your next question: For many types of voices, what is the ratio of PEP-to-average power during a modulation peak in a single-sideband phone signal?

A. Approximately 100 to 1.
B. Approximately 2.5 to 1.
C. Approximately 25 to 1.
D. Approximately 1.0 to 1.

Block 65

The correct answer is B.

Here is your next question: How can a double-sideband phone signal be produced?

A. By modulating the plate supply voltage to a class-C amplifier.
B. By using a phase detector, oscillator and filter in a feedback loop.
C. By varying the voltage to the varactor in an oscillator circuit.
D. By using a reactance modulator on an oscillator.

Block 66

The correct answer is A.

Here is your next question: What type of emission is produced when a frequency modulated transmitter is modulated by a television signal?

A. F3C.
B. F3F.
C. A3C.
D. A3F.

Block 67

The correct answer is B.

Here is your next question: What type of emission is produced when an amplitude modulated transmitter is modulated by a television signal?

A. F3C.
B. A3C.
C. A3F.
D. F3F.

Block 68

The correct answer is C.

Here is your next question: What type of emission is produced when a frequency modulated transmitter is modulated by a facsimile signal?

A. A3F.
B. F3F.
C. A3C.
D. F3C.

Block 69

The correct answer is D.

Here is your next question: What is facsimile?

A. The transmission of moving pictures by electrical means.
B. The transmission of printed pictures by electrical means.
C. The transmission of a pattern of printed characters designed to form a picture.
D. The transmission of tone-modulated telegraphy.

Block 70

The correct answer is B.

Here is your next question: What is emission A3C?

A. Slow-scan TV.
B. ATV.
C. RTTY.
D. Facsimile.

Block 71

The correct answer is D.

Here is your next question: What is the term for the time required for the capacitor in an RC circuit to be charged to 63.2% of the supply voltage?

A. A time factor of one.
B. One exponential period.
C. One time constant.
D. An exponential rate of one.

Block 72

The correct answer is C.

Here is your next question: What is the meaning of the term *time constant* of an RC circuit?

A. The time required to charge the capacitor in the circuit to 63.2% of the supply voltage.
B. The time required to charge the capacitor in the circuit to 63.2% of the supply current.
C. The time required to charge the capacitor in the circuit to 36.8% of the supply current.
D. The time required to charge the capacitor in the circuit to 36.8% of the supply voltage.

Block 73

The correct answer is A.

Here is your next question: What is an optical shaft encoder?

A. An array of optocouplers whose propagation velocity is controlled by a rotating wheel.
B. An array of optocouplers whose propagation velocity is controlled by a stationary wheel.
C. An array of optocouplers whose light transmission path is controlled by a rotating wheel.
D. An array of optocouplers chopped by a stationary wheel.

Block 74

The correct answer is C.

Here is your next question: What is the effective radiated power of a repeater with 200-W transmitter power output, 4-dB feedline loss, 3-dB duplexer and circulator loss, and 6-dB antenna gain?

A. 159 W, assuming the antenna gain is referenced to a half-wave dipole.
B. 632 W, assuming the antenna gain is referenced to a half-wave dipole.
C. 63.2 W, assuming the antenna gain is referenced to a half-wave dipole.
D. 252 W, assuming the antenna gain is referenced to a half-wave dipole.

Block 75

The correct answer is A.

Receivers

Block 1

Here is your first question: What is the result of cross-modulation?

A. Inverted sidebands in the final stage of the amplifier.
B. The modulation of an unwanted signal is heard on the desired signal.
C. Receiver quieting.
D. A decrease in modulation level of transmitted signals.

Block 2

The correct answer is B.

Here is your next question: What is the term used to refer to the condition where the signal from a very strong station are superimposed on other signals being received?

A. Capture effect.
B. Receiver quieting.
C. Cross-modulation interference.
D. Intermodulation distortion.

Block 3

The correct answer is C.

Here is your next question: How can receiver desensitizing be reduced?

A. Increase the receiver bandwidth.
B. Decrease the receiver squelch gain.
C. Increase the transmitter audio gain.
D. Ensure good RF shielding between the transmitter and receiver.

Block 4

The correct answer is D.

Here is your next question: What is the term used to refer to the reduction of receiver gain caused by the signals of a nearby station transmitting in the same frequency band?

A. Squelch gain rollback.
B. Cross-modulation interference.
C. Quieting.
D. Desensitizing.

Block 5

The correct answer is D.

Here is your next question: The band of frequencies least susceptible to atmospheric noise and interference is:

A. 300 to 3000 MHz.
B. 3 to 30 MHz.
C. 300 to 3000 kHz.
D. 30 to 300 kHz.

Block 6

The correct answer is A.

Here is your next question: Which emission modes suffer the most from selective fading?

A. SSTV and CW.
B. SSB and image.
C. FM and double sideband AM.
D. CW and SSB.

Block 7

The correct answer is C.

Here is your next question: What is the propagation effect called when phase differences between radio wave components of the same transmission are experienced at the recovery station?

A. Phase shift.
B. Selective fading.
C. Diversity reception.
D. Faraday rotation.

Block 8

The correct answer is B.

Here is your next question: Two-way communications with both stations operating on the same frequency is:

A. multiplex.
B. simplex.
C. duplex.
D. radiotelephone.

Block 9

The correct answer is B.

Here is your next question: Which VHF channel is used only for digital selective calling?

A. Channel 6.
B. Channel 22A.
C. Channel 16.
D. Channel 70

Block 10

The correct answer is D.

Here is your next question: What is the international VHF digital selective calling channel?

A. 500 kHz.
B. 156.525 MHz.
C. 156.35 MHz.
D. 2182 kHz

Block 11

The correct answer is B.

Here is your next question: How can cross-modulation in a receiver be reduced?

A. By adjusting the pass-band tuning.
B. By increasing the receiver's RF gain while decreasing the AF gain.
C. By using a filter at the receiver.
D. By installing a filter at the receiver.

Block 12

The correct answer is D.

Here is your next question: What is cross-modulation interference?

A. Modulation from an unwanted signal is heard in addition to the desired signal.
B. Harmonic distortion of the transmitted signal.
C. Interference caused by audio rectification in the receiver pre-amp.
D. Interference between two transmitters of different modulation type.

Block 13

The correct answer is A.

Here is your next question: What is the term used to refer to a reduction in receiver sensitivity caused by unwanted high-level adjacent channel signals?

A. Overloading.
B. Desensitizing.
C. Quieting.
D. Intermodulation distortion.

Block 14

The correct answer is B.

Here is your next question: What is receiver desensitizing?

A. A reduction in receiver sensitivity when the AF gain control is turned down.
B. A reduction in receiver sensitivity because of a strong signal on a nearby frequency.
C. A burst of noise when the squelch is set too high.
D. A burst of noise when the squelch is set too low.

Block 15

The correct answer is B.

Here is your next question: How does the bandwidth of the transmitted signal affect selective fading?

A. The receiver bandwidth determines the selective fading effect.
B. It is equally pronounced at both narrow and wide bandwidths.
C. It is more pronounced at narrow bandwidths.
D. It is more pronounced at wide bandwidths.

Block 16

The correct answer is D.

Here is your next question: What is the major cause of selective fading?

A. Phase differences between radio wave components of the same transmission, as experienced at the receiving station.
B. Time differences between the receiving and transmitting stations.
C. Large changes in the height of the ionosphere, as experienced at the receiving station.
D. Small changes in beam heading at the receiving station.

Block 17

The correct answer is A.

Here is your next question: What is a selective fading effect?

A. A fading effect caused by time differences between the receiving and transmitting stations.
B. A fading effect caused by large changes in the height of the ionosphere, as experienced at the receiving station.
C. A fading effect caused by phase differences between radio wave components of the same transmission, as experienced at the receiving station.
D. A fading effect caused by small changes in beam heading at the receiving station.

Block 18

The correct answer is C.

Here is your next question: As an alternative to keeping watch on a working frequency in the band 1600 to 4000 kHz, an operator must tune station receiver to monitor 2182 kHz:

A. during the silence periods each hour.
B. during daytime hours of service.
C. during distress calls only.
D. at all times.

Block 19

The correct answer is D.

Here is your next question: What channel must compulsorily equipped vessels monitor at all times in the open sea?

A. Channel 6, 156.3 MHz.
B. Channel 22A, 157.1 MHz.
C. Channel 16, 156.8 MHz.
D. Channel 8, 156.4 MHz.

Block 20

The correct answer is C.

Here is your next question: In what frequencies does the Communications Act require radio watches by compulsory radiotelephone stations?

A. Watches are required on 2182 kHz and 156.800 MHz.
B. On all frequencies between 405 to 535 kHz, 1605 to 3500 kHz and 156 to 162 MHz.
C. Continuous watch is required on 2182 kHz only.
D. Watches are required on 500 kHz and 2182 kHz.

Block 21

The correct answer is A.

Here is your next question: What is the output impedance of a theoretically ideal op-amp?

A. 1000 Ω.
B. 100 Ω.
C. Very high.
D. Very low.

Block 22

The correct answer is B.

Here is your next question: What technique can be used to construct low-cost, high-performance crystal-lattice filters?

A. Etching and grinding.
B. Etching and splitting.
C. Tumbling and grinding.
D. Splitting and tumbling.

Block 23

The correct answer is A.

Here is your next question: What would be the bandwidth of a good crystal-lattice bandpass filter for a double-sideband phone emission?

A. 15 kHz at –6 dB.
B. 6 kHz at –6 dB.
C. 500 Hz at –6 dB.
D. 1 kHz at –6 dB.

Block 24

The correct answer is C.

Here is your next question: Auto interference to radio reception can be eliminated by:

A. installing two copper-braid ground strips.
B. installing resistors in series with the spark plugs.
C. installing capacitive spark plugs.
D. installing resistive spark plugs.

Block 25

The correct answer is B.

Here is your next question: In radio circuits, the component most apt to break down is the:

A. wiring.
B. transformer.
C. crystal.
D. resistor.

Block 26

The correct answer is C.

Here is your next question: What causes receiver desensitizing?

A. Squelch gain adjusted too low.
B. The presence of a strong signal on a nearby frequency.
C. Squelch gain adjusted too high.
D. Audio gain adjusted too low.

Block 27

The correct answer is B.

Here is your next question: What is a major cause of atmospheric static?

A. Meteor showers.
B. Airplanes.
C. Thunderstorms.
D. Sunspots.

Block 28

The correct answer is A.

Here is your next question: How can ferrite beads be used to suppress ignition noise?

A. Install them in the antenna lead to the radio.
B. Install them in the primary and secondary ignition leads.
C. Install them between the starter solenoid and the starter motor.
D. Install them in the resistive high-voltage cable every two years.

Block 29

The correct answer is D.

Here is your next question: What is one of the most significant problems you might encounter when you try to receive signals with a mobile station?

A. Mechanical vibrations.
B. Radar interference.
C. Doppler shift.
D. Ignition noise.

Block 30

The correct answer is D.

Here is your next question: What is the term used to refer to the reception blockage of one FM-phone signal by another FM-phone signal?

A. Frequency discrimination.
B. Capture effect.
C. Cross-modulation interference.
D. Desensitization.

Block 31

The correct answer is B.

Here is your next question: What is the input impedance of a theoretically ideal op-amp?

A. Very high.
B. Very low.
C. 1000 Ω.
D. 100 Ω.

Block 32

The correct answer is A.

Here is your next question: What is a crystal-lattice filter?

A. A filter with narrow bandwidth and steep skirts made using quartz crystals.
B. A filter with infinitely wide and shallow skirts made using quartz crystals.
C. An audio filter made with four quartz crystals at 1-kHz intervals.
D. A power supply filter made with crisscrossed quartz crystals.

Block 33

The correct answer is B.

Here is your next question: What would be the bandwidth of a good crystal-lattice bandpass filter for a single-sideband phone emission?

A. 15 kHz at –6 dB.
B. 500 Hz at –6 dB.
C. 2.1 kHz at –6 dB.
D. 6 kHz at –6 dB

Block 34

The correct answer is D.

Here is your next question: Motorboating (low-frequency oscillations) in an amplifier can be stopped by:

A. grounding the plate.
B. connecting a capacitor between the B+ lead and ground.
C. by passing the screen grid resistor with a 0.1-μF capacitor.
D. grounding the screen grid.

Block 35

The correct answer is D.

Here is your next question: How can alternator whine be minimized?

A. By installing filter capacitors in series with the dc power lead.
B. By installing a high-pass filter in series with the radio's dc power lead to the vehicle's electrical system.

C. By connecting the radio's power leads to the battery by the shortest possible path.

D. By connecting the radio's power leads to the battery by the longest possible path.

Block 36

The correct answer is C.

Here is your next question: How can you determine if a line-noise interference problem is being generated within a building?

A. Observe the power-line voltage on a spectrum analyzer.
B. Turn off the main circuit breaker and listen on a battery-operated radio.
C. Observe the ac waveform on an oscilloscope.
D. Check the power-line voltage with a time-domain reflectometer.

Block 37

The correct answer is C.

Here is your next question: How can conducted and radiated noise caused by an alternator be suppressed?

A. By connecting the radio's power leads directly to the battery and by installing coaxial capacitors in the alternator leads.
B. By installing a high-pass filter in series with the radio's power lead to the vehicle's electrical system and by installing a low-pass filter in parallel with the field lead.
C. By connecting the radio's power leads to the battery by the longest possible path and by installing a blocking capacitor in series with the positive lead.
D. By installing filter capacitors in series with the dc power lead and by installing a blocking capacitor in the field lead.

Block 38

The correct answer is B.

Here is your next question: What is the proper procedure for suppressing electrical noise in a mobile station?

A. Install filter capacitors in series with all dc wiring.
B. Apply anti-static spray liberally to all nonmetallic surfaces.
C. Insulate all plane sheet-metal surfaces from each other.
D. Apply shielding and filtering where necessary.

Block 39

The correct answer is D.

Here is your next question: With which emission type is the capture-effect most pronounced?

A. CW.
B. AM.

C. SSB.
D. FM.

Block 40

The correct answer is B.

Here is your next question: What is the capture effect?

A. The weakest signal received is the only demodulated signal.
B. The loudest signal received is the only demodulated signal.
C. All signals on a frequency are demodulated by an AM receiver.
D. All signals on a frequency are demodulated by an FM receiver.

Block 41

The correct answer is B.

Here is your next question: Where is the noise generated that primarily determines the signal-to-noise ratio in a VHF (150 MHz) marine-band receiver?

A. In the ionosphere.
B. In the atmosphere.
C. Man-made noise.
D. In the receiver front end.

Block 42

The correct answer is D.

Here is your next question: What is meant by the term *noise figure* of a communications receiver?

A. The ability of a receiver to reject unwanted signals at frequencies close to the desired one.
B. The level of noise generated in the front end and succeeding stages of a receiver.
C. The relative strength of a received signal 3 kHz removed from the carrier frequency.
D. The level of noise entering the receiver from the antenna.

Block 43

The correct answer is B.

Here is your next question: What type of problems are caused by poor dynamic range in a communications receiver?

A. Oscillator instability and severe audio distortion of all but the strongest received signals.
B. Cross-modulation of the desired signal and insufficient audio power to operate the speaker.
C. Oscillator instability requiring frequency retuning, and loss of ability to recover the opposite sideband, should it be transmitted.
D. Cross-modulation of the desired signal and desensitization from strong adjacent signals.

Block 44

The correct answer is D.

Here is your next question: What is meant by the dynamic range of a communications receiver?

A. The difference between the lowest-frequency signal and the highest-frequency signal detectable without moving the tuning knob.
B. The ratio between the minimum discernible signal and the largest tolerable signal without causing audible distortion products.
C. The maximum possible undistorted audio output of the receiver, referenced to one milliwatt.
D. The number of kHz between the lowest and the highest frequency to which the receiver can be tuned.

Block 45

The correct answer is B.

Here is your next question: What degree of selectivity is desirable in the IF circuitry of an FM-phone receiver?

A. 15 kHz.
B. 4.2 kHz.
C. 2.4 kHz.
D. 1 kHz.

Block 46

The correct answer is A.

Here is your next question: What is an undesirable effect of using too wide a filter bandwidth in the IF section of a receiver?

A. Filter ringing.
B. Thermal-noise distortion.
C. Undesired signals will reach the audio stage.
D. Output-offset overshoot.

Block 47

The correct answer is C.

Here is your next question: A receiver selectivity of 10 kHz in the IF circuitry is optimum for what type of signals?

A. FSK (frequency shift keying) RTTY.
B. CW.
C. Double-sideband AM.
D. SSB voice.

Block 48

The correct answer is C.

Here is your next question: A receiver selectivity of 2.4 kHz in the IF circuitry is optimum for what type of signals?

A. FSK RTTY.
B. Double-sideband AM voice.
C. SSB voice.
D. CW.

Block 49

The correct answer is C.

Here is your next question: What is the theoretical minimum noise floor of a receiver with a 400-Hz bandwidth?

A. to 180 dBm.
B. to 174 dBm.
C. to 148 dBm.
D. to 141 dBm.

Block 50

The correct answer is C.

Here is your next question: What two factors determine the sensitivity of a receiver?

A. Bandwidth and noise figure.
B. Intermodulation distortion and dynamic range.
C. Cost and availability.
D. Dynamic range and third-order intercept.

Block 51

The correct answer is A.

Here is your next question: What is one major advantage of CMOS over other devices?

A. Ease of circuit design.
B. Low cost.
C. Low current consumption.
D. Small size.

Block 52

The correct answer is A. That is the answer given by the FCC.

Here is your next question: Why do circuits containing TTL devices have several bypass capacitors per printed circuit board?

A. To prevent switching transients from appearing on the supply line.
B. To filter out switching harmonics.
C. To keep the switching noise within the circuit, thus eliminating RFI.
D. To prevent RFI (radio frequency interference) to receivers.

Block 53

The correct answer is D.

Here is your next question: What level of input voltage is high in a TTL device operating with a 5-V power supply?

A. 5.0 to 2.0 V.
B. 1.0 to 1.5 V.
C. 1.5 to 3.0 V.
D. 2.0 to 5.5 V.

Block 54

The correct answer is B.

Here is your next question: What is the recommended power supply voltage for TTL series integrated circuits?

A. 13.60 V.
B. 5.00 V.
C. 50.00 V.
D. 12.00 V.

Block 55

The correct answer is B.

Here is your next question: A circuit compares the output from a voltage-controlled oscillator and a frequency standard. The difference between the two frequencies produces an error voltage that changes the voltage-controlled oscillator frequency. What is the name of the circuit?

A. A variable frequency oscillator.
B. A differential voltage amplifier.
C. A phase-locked loop.
D. A doubly balanced mixer.

Block 56

The correct answer is B.

Here is your next question: What is a phase-locked loop circuit?

A. An electronic servo loop consisting of a phase detector, a low-pass filter, and voltage-controlled oscillator.
B. An electronic circuit consisting of a precision push-pull amplifier with a differential input.
C. An electronic circuit also known as a *monostable multivibrator*.
D. An electronic servo loop consisting of a ratio detector, reactance modulator, and voltage-controlled oscillator.

Block 57

The correct answer is A.

Here is your next question: Which stage of a receiver primarily establishes its noise figure?

A. The local oscillator.
B. The RF stage.
C. The IF strip.
D. The audio stage.

Block 58

The correct answer is B.

Here is your next question: The ability of a communications receiver to perform well in the presence of strong signals outside the band of interest is indicated by what parameter?

A. Audio output.
B. Signal-to-noise ratio.
C. Blocking dynamic range.
D. Noise figure.

Block 59

The correct answer is C.

Here is your next question: What is the term for the ratio between the largest tolerable receiver input signal and the minimum discernible signal?

A. Dynamic range.
B. Noise figure.
C. Noise floor.
D. Intermodulation distortion.

Block 60

The correct answer is A.

Here is your next question: How can selectivity be achieved in the IF circuitry of a communications receiver?

A. Incorporate a high-Q filter.
B. Remove AGC action from the IF stage and confine it to the audio stage only.
C. Replace the standard JFET mixer with a bipolar transistor followed by a capacitor of the proper value.
D. Incorporate a means of varying the supply voltage to the local oscillator circuitry.

Block 61

The correct answer is A.

Here is your next question: How should the filter bandwidth of a receiver IF section compare with the bandwidth of a received signal?

A. Filter bandwidth should be approximately four times the received-signal bandwidth.

B. Filter bandwidth should be approximately two times the received-signal bandwidth.
C. Filter bandwidth should be approximately half the received-signal bandwidth.
D. Filter bandwidth should be slightly greater than the received-signal bandwidth.

Block 62

The correct answer is D.

Here is your next question: What degree of selectivity is desirable in the IF circuitry of a single-sideband phone receiver?

A. 4.8 kHz.
B. 4.2 kHz.
C. 2.4 kHz.
D. 1 kHz

Block 63

The correct answer is C.

Here is your next question: What occurs during CW reception if too narrow a filter bandwidth is used in the IF stage of a receiver?

A. Filter ringing.
B. Cross-modulation distortion.
C. Output-offset overshoot.
D. Undesired signals will reach the audio stage.

Block 64

The correct answer is A.

Here is your next question: How can selectivity be achieved in the front-end circuitry of a communications receiver?

A. By using an additional IF amplifier stage.
B. By using an additional RF amplifier stage.
C. By using a preselector.
D. By using an audio filter.

Block 65

The correct answer is C.

Here is your next question: What is the limiting condition for sensitivity in a communications receiver?

A. The input impedance to the detector.
B. The two-tone intermodulation distortion.
C. The power-supply output ripple.
D. The noise floor of the receiver.

Block 66

The correct answer is D.

Here is your next question: Why do CMOS digital integrated circuits have high immunity to noise on the input signal or power supply?

A. Input signals are stronger.
B. The input switching threshold is about one-half the power supply voltage.
C. The input switching threshold is about two times the power supply voltage.
D. Larger bypass capacitors are used in CMOS circuit design.

Block 67

The correct answer is C.

Here is your next question: What is a CMOS IC?

A. A chip with only bipolar transistors.
B. A chip with only N-channel transistors.
C. A chip with P-channel and N-channel transistors.
D. A chip with only P-channel transistors.

Block 68

The correct answer is B.

Here is your next question: What level of input voltage is low in a TTL device operating with a 5-V power supply?

A. 0.8 to 0.4 V.
B. 0.6 to 0.8 V.
C. 2.0 to 5.5 V.
D. 2.0 to 5.5 V.

Block 69

The correct answer is B.

Here is your next question: What logic state do the inputs of a TTL device assume if they are left open?

A. Open inputs on a TTL device are ignored.
B. The device becomes randomized and will not provide consistent high or low logic states.
C. A low logic state.
D. A high logic state.

Block 70

The correct answer is C.

Here is your next question: What do the initials TTL stand for?

A. Emitter-coupled logic.

B. Diode-transistor logic.
C. Transistor-transistor logic.
D. Resistor-transistor logic.

Block 71

The correct answer is C.

Here is your next question: What functions are performed by a phase-locked loop?

A. Frequency synthesis, FM demodulation.
B. Photovoltaic conversion, optical coupling.
C. Comparison of two digital input signals, digital pulse counter.
D. Wideband AF and RF power amplification.

Block 72

The correct answer is A.

7
CHAPTER

Associate-level and journeyman-level CET practice test

Basic equations that you are expected to know in the mathematics section are: Ohm's law, power, impedance, time constant, and frequency. In other words, know the following:

$$I = V/R$$
$$I = V/Z \text{ (purely resistive circuit)}$$
$$(Z \text{ substitutes for } R \text{ in purely resistive circuits)}$$

$$P_{dc} = I^2R = V \times I = V^2/R$$
$$P_{ac} = V \times I \times \text{Cos } 0$$
$$Z = \sqrt{R^2 + (X_L - X_C)^2}$$

$$T = RC = L/R \text{ (time constant)}$$
$$T = 1/f \text{ or } f = 1/T \text{ (period of an ac wave)}$$

where:
  T is the period, and
  f is the frequency of a wave

Also, know the resistor color code, including the tolerances and be able to calculate the maximum or minimum allowable resistance for a resistor to be within tolerance.

Know how to calculate the sweep time or frequency from an oscilloscope display. Example: If Fig. A shows a waveform on a triggered sweep oscilloscope, and the sweep is calibrated for 5 μs, what is the frequency of the waveform? Answer: There was one complete cycle in 2.5 μs and the frequency is:

$$f = \frac{1}{T}$$

$$= \frac{1}{2.5 \times 10^{-6}} = 400{,}000 \text{ Hz}$$

$$= 400 \text{ kHz}$$

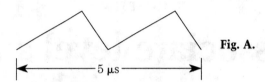

Fig. A.

5 μs

Remember that voltages across reactive components (inductors and capacitors) in ac circuits are in quadrature with voltages across resistors. So, in the circuit of Fig. B, the generator voltage is not 50 V ($V = 36$ V).

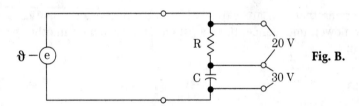

Fig. B.

1. Refer to Fig. 1-1. What is the voltage at point A with respect to the voltage at point B?

    A. 1.0 V.
    B. 1.6 V.
    C. 3.3 V.
    D. 4.0 V.

2. Refer to Fig. 1-1. How much current is flowing through the center resistor?

    A. about 3.5 mA.
    B. about 7.0 mA.
    C. about 35 mA.
    D. about 70 mA.

3. Could you use a 5-V power supply that is rated at 50 mW to operate the circuit of Fig. 1-2? (Assume that there is no outboard circuitry except for the power supply.)

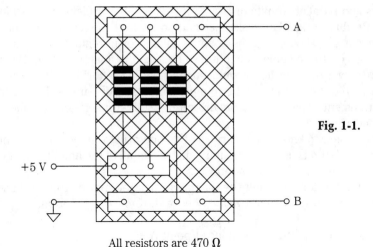

Fig. 1-1.

All resistors are 470 Ω

$R = 50\ \Omega$

$X_C = 20\ \Omega$

Fig. 1-2.

A. Yes, because the power dissipated by the three resistors is less than 50 mW.
B. No, because the power dissipated by the three resistors is greater than 50 mW.

4. Assume that the power supply in the circuit of Fig. 1-1 is disconnected. The resistors are each color coded yellow, violet, brown, and gold. What is the lowest allowable resistance between terminals A and B?

A. About 670 Ω.
B. About 376 Ω.
C. About 423 Ω.
D. About 447 Ω.

5. Consider the following statement about Fig. 1-2: The impedance between terminals A and B is:

$$50\ \Omega + 20\ \Omega = 70\ \Omega$$

A. The statement is correct.
B. The statement is not correct.

All of the amplifying devices (tubes, bipolar transistors, and field-effect transistors) require a dc voltage for their operation. This means that a technician must be able to analyze dc circuits, trace dc current paths, and calculate voltage, current, resistance, and power.

Series and parallel circuits of resistors, and series-parallel combinations might be given. Be able to calculate resistance and power in these resistor circuits.

You might be asked questions about regulated or unregulated power supplies, but questions on these subjects might also be asked in a later section. Be sure that you know the various rectifier layouts, such as half wave, bridge, doublers, etc. Remember that rectifier diodes can be connected in series or parallel. Equalizing resistors and transient bypass capacitors are often used with series or parallel rectifier diode connections.

It is necessary to know the polarities of dc operating voltage for all of the amplifying devices. Methods used for obtaining dc bias for all of the amplifying devices are also subjects that you might encounter.

The condition for obtaining the maximum possible power from a dc source is called the *Maximum Power Transfer Theorem*. Know this theorem.

6. In the circuit of Fig. 2-1, the voltage at point A is:

   A. –5 V.
   B. 0 V.
   C. +5 V.
   D. –11 V.

   Note: In questions like this, it is always correct to assume that voltages are given with respect to ground.

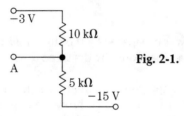

Fig. 2-1.

7. Refer to Fig. 2-2. In order for this component to be properly biased for class-A operation, the gate should be:

   A. positive with respect to common.
   B. negative with respect to common.

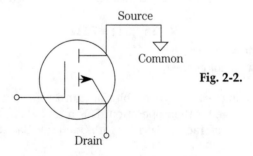

Fig. 2-2.

8. For the component of Fig. 2-2, to be biased for class-A operation, its source must be:

    A. positive with respect to the gate.

    B. negative with respect to the gate.

9. Regarding the dc supply of Fig. 2-3, the voltage between terminals A and B should be:

    A. less than 110 V.

    B. more than 130 V.

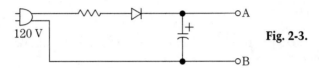

Fig. 2-3.

10. In the circuit of Fig. 2-4, the load resistor (R_L) will receive the maximum possible power from the circuit when it is adjusted to:

    A. 1.25 Ω.

    B. 2.5 Ω.

    C. 5 Ω.

    D. The answer cannot be determined from the information given.

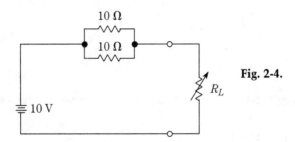

Fig. 2-4.

Although the amplifying devices are operated with dc voltages, the signals are ac. The reactive voltages are in quadrature with the resistive components, and the reactances and resistances are related in an impedance triangle.

Sine waves are the basic waveforms used in the study of ac. Be able to determine peak, RMS, and average values.

The characteristics of series and parallel tuned circuits are important. Remember that parallel tuned circuits can be tuned with resistors.

An understanding of the mathematics of complex variables is not required, so you will not be asked to solve problems in j operators. And, you will not be required to calculate resonant frequencies, inductive or capacitive reactances, or impedance values. Such calculations are important in your training because they help you to understand ac circuits, but they are not part of the CET test.

Be sure that you know the relationships between inductors and capacitors and their reactances. For example, the reactance of an inductor increases with frequency, and the reactance of a capacitor decreases with frequency. These facts are important in understanding the effect of a frequency change in an ac circuit.

11. In any given R-L circuit, which of the following has the highest value?

    A. Real power.

    B. Apparent power.

    C. VARS (reactive volt-amperes).

    D. It is not possible to tell from the information given.

12. Figure 3-1 shows two resonant curves for a series RLC circuit. Which curve is for a higher value of R?

    A. The curve marked x.

    B. The curve marked y.

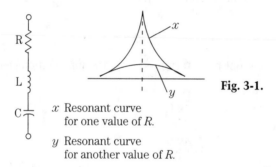

Fig. 3-1.

x Resonant curve for one value of R.

y Resonant curve for another value of R.

13. Is the following statement true? In a purely inductive circuit, there is no power dissipated.

    A. The statement is true.

    B. The statement is false.

14. Figure 3-2 shows two resonant curves for a parallel RLC circuit. Which curve is for a higher value of R?

    A. The curve marked x.

    B. The curve marked y.

15. In the circuit of Fig. 3-3, the voltmeter reading will:

    A. be equal to the generator voltage minus the voltage across the capacitor.

    B. not be affected by a change in generator frequency.

    C. decrease if the generator frequency is increased.

    D. increase if the generator frequency is increased.

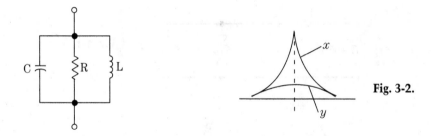

x—Resonant curve for one value of R.
y—Resonant curve for another value of R.

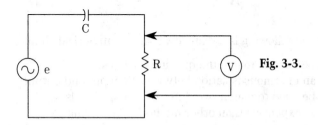

In this practice test, Sections IV and V are titled as follows:

Section IV: Transistors and Semiconductors
Section V: Electronic Components and Circuits

In this section, you can expect to find questions on bipolar transistors, field-effect transistors, UJTs, semiconductor diodes, thyristors, LEDs, and light-activated components.

In a few cases, you will find a very basic circuit, such as the one in Fig. 4-1, but the answer to the related question depends upon a knowledge of what the component is used for.

In this section, you might also get questions on the dc voltages or bias circuits used for operating the electronic devices. If you are not sure about the dc operating voltages for these semiconductor devices, you should take some time to review them before taking the actual CET test.

16. The input voltage to the circuit of Fig. 4-1 is a sine wave of 5 V RMS, and the two 5-V zeners are connected back to back. Which of the waveforms in Fig. 4-2 is correct for the output across terminals X and Y?

    A. The one marked A.
    B. The one marked B.
    C. The one marked C.
    D. The one marked D.

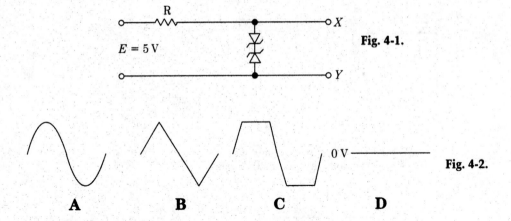

Fig. 4-1.

Fig. 4-2.

A   B   C   D

17. Which of the following is not an advantage of an optical coupler?

    A. It can be used to match unequal impedances.
    B. It has an enormous isolation between the input and output.
    C. It can be used to match two different voltage levels.
    D. It is less expensive than other methods of coupling.

18. Which of the following types of diodes is normally operated with reverse bias?

    A. A varactor diode.
    B. An LED.
    C. A rectifier diode.
    D. A tunnel diode.

19. Which of the following types of diodes can be used as a very fast-acting switch?

    A. A varactor diode.
    B. An LED.
    C. A rectifier diode.
    D. A tunnel diode.

20. Which of the following components is most nearly like a triac?

    A. Two rectifier diodes connected back to back.
    B. Two thyratrons connected back to back.
    C. Two variacs in series.
    D. Two triodes in a cascode connection.

21. In the circuit of Fig. 4-3, the switch is closed and then opened. The light is on when the switch is closed.

    A. The light will stay on after the switch is opened.
    B. The light will not stay on after the switch is opened.

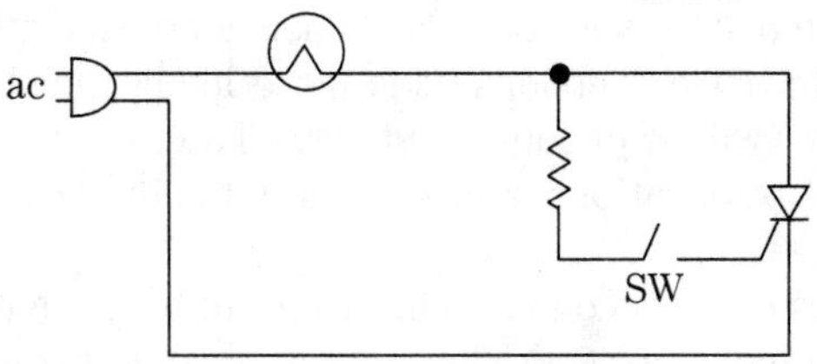

ac

Fig. 4-3.

SW

22. If the plates of a capacitor are moved closer together, the capacitance will:

    A. increase.
    B. decrease.

23. Which of the following is an example of a breakover diode?

    A. Magnetron.
    B. LED (3) diac.
    C. Light-activated diode.

24. The symbol in Fig. 4-4 represents:

    A. a P-channel enhancement-type MOSFET.
    B. an N-channel enhancement-type MOSFET.
    C. a P-channel depletion-type MOSFET.
    D. an N-channel depletion-type MOSFET.

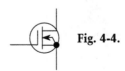

Fig. 4-4.

25. The emitter-base voltage of a germanium transistor in a class-A amplifier circuit should be about:

    A. 0.2 V.
    B. 0.8 V.
    C. 5 V.
    D. 9 V.

The emphasis in this section is on circuits that are used in a wide variety of systems. Remember that all prospective CETs must take the Associate-Level test, so the circuits in Section V are not necessarily found only in television sets or other consumer products.

There are some fundamental circuits that consistently occur in this section. Rectifiers, amplifiers, coupling circuits between amplifiers, oscillators, and operational amplifiers are the most common.

In early versions of the test, it was common to use neon oscillators as examples of relaxation oscillators. In newer versions, it is more likely that UJTs, SCSs, and solid-state breakover diodes will be encountered in oscillator circuits. These circuits are basically time-constant applications, and you should know how their frequency is determined.

Technicians sometimes fail to recognize a basic circuit because it is not drawn in a standard fashion. You should never try to learn circuitry strictly on the basis of the way it is drawn. The values of components are often a tipoff on the type of circuit, and the application of the circuit. Practice tracing dc and ac (signal) paths as part of your procedure for analyzing circuits.

In newer versions of the Associate-Level test, you might find some basic logic circuitry in this section.

26. The ripple frequency of a full-wave rectifier that is operated with a 60-Hz power line input is:

    A. 120 Hz.
    B. 90 Hz.
    C. 60 Hz.
    D. 30 Hz.

27. Which of the following is a likely purpose of R and C in the coupling circuit of Fig. 5-1?

    A. False bass tone control.
    B. High-frequency compensation.
    C. Decoupling filter.
    D. De-emphasis.

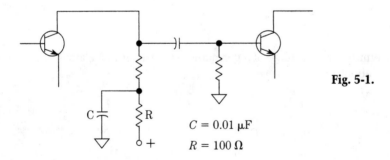

Fig. 5-1.

$C = 0.01\ \mu F$

$R = 100\ \Omega$

28. Which of the following components is sometimes used as a parasitic suppressor?

    A. A bead ledge.
    B. A four-layer diode.
    C. A ferrite bead.
    D. A thermistor.

29. The voltage gain of the op-amp circuit in Fig. 5-2 is:

   A. R_a/R_b.
   B. R_b/R_a.
   C. $R_L/R_a + R_b$.
   D. $R_a + R_b/R_L$.

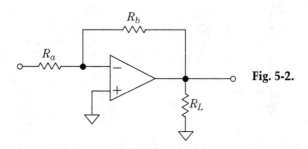

Fig. 5-2.

30. The power supply connection for the op-amp circuit in Fig. 5-2 is:

   A. across the + and − terminals.
   B. between the + terminal and the output terminal.
   C. between the output terminal and ground.
   D. not shown.

31. Which of the following circuit configurations has a high-input impedance and low-output impedance?

   A. Common emitter.
   B. Grounded grid.
   C. Source follower.
   D. None of these.

32. In which of the following types of tube bias could a destructive plate current result from a loss of signal?

   A. Contact bias.
   B. Cathode bias.
   C. Battery bias.
   D. Grid-leak bias.

33. A bootstrap circuit is used with a bipolar transistor amplifier circuit to:

   A. increase the gain.
   B. reduce power supply loading.
   C. increase the input impedance.
   D. make the gain independent of power supply voltage fluctuations.

34. In the relaxation oscillator circuit of Fig. 5-3, increasing the resistance of R will:

    A. increase the output frequency.
    B. decrease the output frequency.

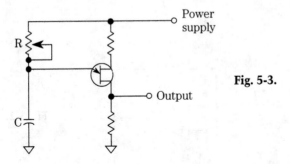

Fig. 5-3.

35. Refer to Fig. 5-3. Which of the waveforms of Fig. 5-4 would you expect to observe with an oscilloscope at the output terminal?

    A. The one shown in A.
    B. The one shown in B.
    C. The one shown in C.
    D. The one shown in D.

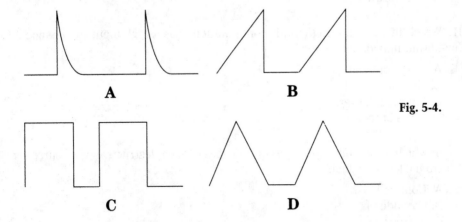

Fig. 5-4.

The questions in this section deal with the instruments used for making tests and measurements in electronics. Of special interest are the instruments used for troubleshooting. You should know the construction features of voltmeters and ammeters and the advantages of taut-band meter movements over the jeweled type.

You are likely to find a number of questions on the accuracy of tests and measurements and how this accuracy is affected by the circuit under test.

Oscilloscopes are important instruments, so you should have a clear understanding of how oscilloscopes work. Be able to compare the triggered sweep and recurrent sweep scopes, and how the scopes are used for measuring voltage, current, time, and phase differences.

In a few cases, you might encounter questions on the use of instruments. For example, you might be asked how an ammeter or voltmeter is connected for making a measurement.

Auxiliary equipment, such as scope probes and electronic switches, might also be the subject of questions in this section.

36. An oscilloscope is displaying a sine wave that has an average value of 6.36 V. If the vertical input of the oscilloscope is calibrated for 10 volts per inch—the display should be:

    A. 0.6 inches high.
    B. 1 inch high.
    C. 1.4 inches high.
    D. 2 inches high.

37. Which of the following statements is correct?

    A. An oscilloscope can be used to display current waveforms.
    B. An oscilloscope cannot be used to display current waveforms.

38. The rise time of the waveform in Fig. 6-1:

    A. is 5 ms.
    B. is 4 ms.
    C. is 3 ms.
    D. cannot be determined from the information given.

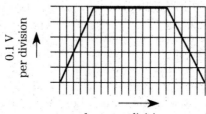

Fig. 6-1.

1 ms per division

39. In comparing a jeweled meter movement with a taut-band meter movement, which of the following statements is correct?

    A. The jeweled meter movement has less friction.
    B. The taut-band meter movement has less friction.

40. It is a common practice to locate thermistors and other measuring sensors in one leg of a Wheatstone bridge. The reason for doing this is that:

    A. it increases the sensitivity of the measurement.
    B. it makes the measurement nearly independent of power-supply voltage variations.

    C. it increases the circuit resistance, as seen by the sensor, much greater.

    D. it permits a maximum power transfer condition.

41. The advantage of a mirrored scale on a voltmeter is that it can be used to:

    A. make a measurement and watch the circuit at the same time.

    B. make sure that no one is approaching the bench when you are making a dangerous measurement.

    C. eliminate the problem of parallax.

    D. None of these answers is correct.

42. A volt-ohm-milliammeter with a taut-band meter movement is normally calibrated to measure RMS values. What voltage would a VOM indicate for the waveform in Fig. 6-2?

    A. 10 V.

    B. 0.7 V.

    C. 0.6 V.

    D. Cannot be determined from the information given.

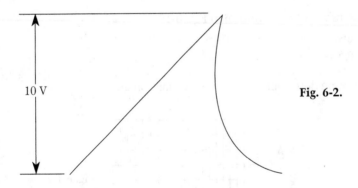

10 V   **Fig. 6-2.**

43. The best way to determine the condition of a LeClanche cell (carbon-zinc dry call) is to:

    A. measure the open-circuit terminal voltage.

    B. measure the short-circuit current.

    C. measure the terminal voltage under normal load.

    D. weigh the cell.

44. Which of the following statements is true?

    A. A high-impedance voltmeter should not be used for making measurements in low-impedance circuits.

    B. A low-impedance voltmeter should not be used for making measurements in high-impedance circuits.

45. A multiplier is used to:

    A. convert a microammeter to a voltmeter.
    B. convert a microammeter to an ammeter.
    C. convert a microammeter to a wattmeter.
    D. increase the light output of an LED.

This section contains questions on voltage and current measurements, resistance measurements, tests and measurements with an oscilloscope, and tests and measurements to evaluate components and devices.

The parameters important to the operation of tubes, bipolar transistors, and FETs are important, and usually measured by testers. For example, a good dynamic tube tester (as opposed to an emission checker) can give accurate numerical values of g_m. As another example, a good transistor checker can provide numerical values. Be sure that you know the meaning of these parameters, and how they relate to the performance of the device in a circuit.

Some of the subjects for this section have been located in other sections. Lissajous patterns, bridge circuits, and the need for sweep generators are examples. The section titles refer to the emphasis, not to the subject of every question.

46. To obtain the lissajous pattern in Fig. 7-1, the frequency of the vertical signal is 480 Hz. What is the frequency of the horizontal signal?

    A. 960 Hz.
    B. 720 Hz.
    C. 320 Hz.
    D. Cannot be determined from the information given.

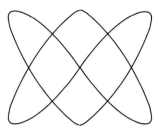

Fig. 7-1.

47. A Wheatstone bridge is used for measuring

    A. capacitance.
    B. inductance.
    C. resistance.
    D. pressure.

48. Some multimeters have a high-power ohms and a low-power ohms switch. The low-power ohms position:

    A. is used to measure resistance values less than 1 Ω.

    B. is used for measuring resistances of resistors having a power rating of less than 1 W.

    C. is used for saving the battery when the meter is used on battery power.

    D. is used for measuring resistance in semiconductor circuits.

49. A lissajous pattern is formed by the input and output sine wave voltages for the amplifier in Fig. 7-2. Ideally, the pattern should be a:

    A. sine wave.

    B. straight line.

    C. circle.

    D. figure eight.

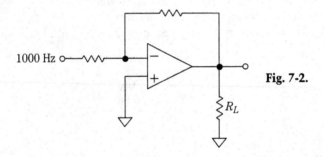

Fig. 7-2.

50. A curve tracer and an oscilloscope are used to obtain the display of Fig. 7-3. The component with this characteristic curve is a:

    A. tunnel diode.

    B. Shockley diode.

    C. hot-carrier diode.

    D. diac.

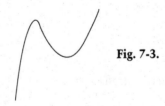

Fig. 7-3.

51. To make the voltage measurement in Fig. 7-4:

    A. the positive lead of the voltmeter goes to point A.

    B. the negative lead of the voltmeter goes to point A.

52. If the emitter is shorted to the base of the transistor in Fig. 7-4, the voltmeter reading will:

    A. increase.
    B. decrease.

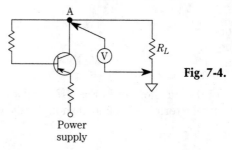

Fig. 7-4.

53. Which of the following is a frequency value?

    A. Beta.
    B. Intrinsic standoff ratio.
    C. Gain-bandwidth product.
    D. None of the factors listed is a frequency value.

54. The oscilloscope display of Fig. 7-5 is obtained by delivering signals to the vertical input and:

    A. external sync input.
    B. horizontal input.
    C. z-axis.
    D. external calibrate input.

Fig. 7-5.

55. In Fig. 7-6, the voltage at point A should be:

    A. positive with respect to common.
    B. negative with respect to common.

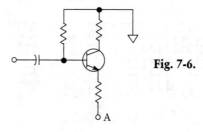

Fig. 7-6.

In order to make voltage measurements in transistor circuits, you should have a reasonably good idea of the polarities of voltages. This is not required, of course, if you have a digital voltmeter with automatic polarity setting.

56. A logic probe is not normally used for measuring a:

    A. pulse condition.
    B. logic 1 condition.
    C. logic 0 condition.
    D. delay condition.

57. Terminals B and C of the transformer in Fig. 7-7 are connected together. An ac voltmeter connected between terminals A and D should indicate:

    A. 48 V.
    B. 24 V.
    C. 12 V.
    D. 0 V.

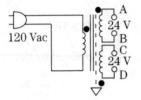

Fig. 7-7.

58. On a transmission line, it is most desirable to have a standing wave ratio of:

    A. 10:1.
    B. 1:1.
    C. 0:1.
    D. 1:0.

59. Figure 7-8 shows a frequency response curve for a voltage amplifier. This type of oscilloscope display is achieved by the use of:

    A. a tone burst generator.
    B. an RF signal generator.
    C. a sweep generator.
    D. a time domain generator.

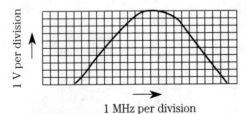

Fig. 7-8.

60. If bandwidth is defined as the range of frequencies between the half-power points, what is the bandwidth of the amplifier having the response curve in Fig. 7-8?

    A. 20 MHz.
    B. 17.5 MHz.
    C. 12.5 MHz.
    D. 10 MHz.

    Section VIII has questions on troubleshooting procedures and on interpretation of test results. Signal injection vs. signal-tracing procedures are other possible subjects of questions.

    You are likely to be asked to determine the amount of voltage, or the amount of current at some point in a circuit. You might also be asked to determine the phase relationships between the input and output signals of a network. Be sure that you understand the square-wave, triangular-wave, and tone-burst tests.

    An important feature of the test is the part that deals with pictorial circuits. You are shown a picture of a circuit and asked to interpret the circuitry. Your experience working with printed circuit boards will give you the background needed for answering these questions. It might be helpful to draw a few circuits from their printed circuit layout.

61. In the circuit of Fig. 8-1, the voltages at points A and B are:

    A. in phase.
    B. 90° out of phase, with voltage at a leading the voltage at B.
    C. 90° out of phase, with the voltage at a lagging the voltage at B.
    D. 180° out of phase.

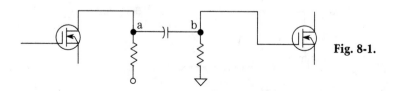

Fig. 8-1.

62. In Fig. 8-2, electrolytic capacitor C is not charged. The voltage at point A is:

    A. 0 V with respect to the voltage at B.
    B. +150 V with respect to the voltage at point B.
    C. negative with respect to the voltage at point B.
    D. None of these answers is correct.

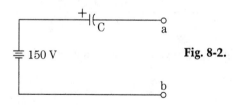

Fig. 8-2.

63. In the circuit of Fig. 8-3, the voltage across terminals A and B is:

    A. 15 V.
    B. 9 V.
    C. 6 V.
    D. 3 V.

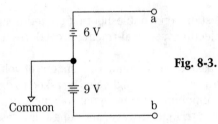

Fig. 8-3.

64. The circuit of Fig. 8-3 is modified by the addition of two resistors as shown in Fig. 8-4. The voltage across terminals A and B is now:

    A. 15 V.
    B. 9 V.
    C. 6 V.
    D. 3 V.

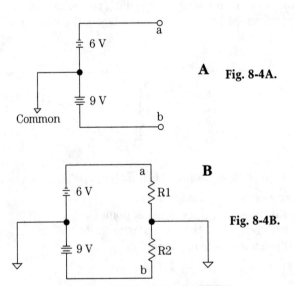

A   Fig. 8-4A.

B

Fig. 8-4B.

65. In the circuit of Fig. 8-5, the voltage across switch SW3 is:

    A. 0 V.
    B. 100 V.

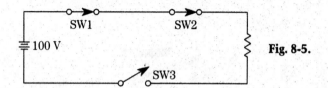

Fig. 8-5.

66. The voltmeter in Fig. 8-6 indicates 1.5 V. This means that the transistor is probably:

   A. cut off.
   B. saturated.

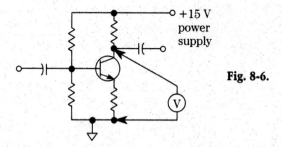

Fig. 8-6.

67. The output of the logic gate in Fig. 8-7:

   A. should be a square wave.
   B. should always be at logic level zero.
   C. should always be at logic level one.
   D. cannot be determined from the information given.

Fig. 8-7.

68. A strong mechanical connection of a wire to a terminal, as shown in Fig. 8-8:

   A. is undesirable and is not required for a good electrical connection.
   B. is necessary before soldering. This ensures a good electrical connection.

69. A certain voltmeter has a rating of 20,000 Ω per V. This means that:

   A. it requires a current of 50 µA to obtain a full-scale deflection of the meter movement.
   B. the voltmeter has an input resistance of 20,000 for each volt being measured. For example, it has an input resistance of 40,000 Ω when measuring 2 V.

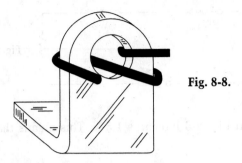

Fig. 8-8.

70. A square wave is delivered to the input of an amplifier, and the output is monitored with an oscilloscope. Figure 8-9 shows the output waveform. The amount of tilt in the top and bottom could be reduced by:

    A. improving the high-frequency response.
    B. improving the low-frequency response.

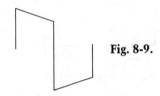

Fig. 8-9.

71. An ohmmeter is used to make measurements on an FET. The resistance values are infinite when the polarities of the ohmmeter leads are as shown in Fig. 8-10A. The values are also infinite when the measurements are as shown in Fig. 8-10B. Which of the following statements is true?

    A. If this is an N-channel JFET, the readings are normal.
    B. If this is a P-channel JFET, the readings are normal.
    C. If this is an N-channel enhancement MOSFET, the readings are normal.
    D. If this is a P-channel depletion MOSFET, the readings are normal.

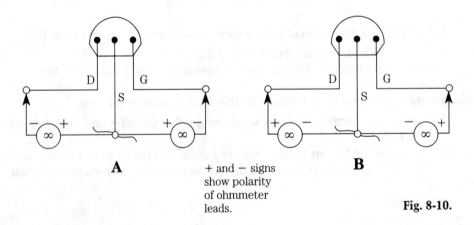

A

+ and − signs
show polarity
of ohmmeter
leads.

B

Fig. 8-10.

72. Figure 8-11 shows a phono amplifier that uses an N-channel JFET (Q1). The power supply should be:

    A. positive.
    B. negative.

73. In Fig. 8-11, gate bias is obtained with:

    A. a power-supply voltage divider.
    B. a resistor between the gate and drain.
    C. a source resistor.
    D. None of these answers is correct.

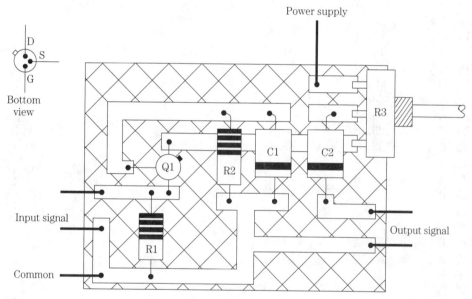

Fig. 8-11.

74. Regarding the circuit in Fig. 8-11, if C is open, the result will be:

    A. a decrease in gain.
    B. no change in gain.
    C. complete loss of output signal.
    D. an increase in gain.

75. What range of output voltages can be obtained by adjusting R2 in Fig. 8-12?

    A. 0 to 0.6 V.
    B. 0 to 0.84 V.
    C. 0 to 3 V.
    D. 0.24 to 0.6 V.

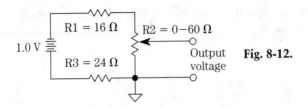

Fig. 8-12.

Answer sheet for
Associate-Level CET Practice Tests

1. C	2. A	3. A
4. D	5. B	6. D
7. A	8. A	9. B
10. C	11. B	12. B
13. A	14. A	15. D
16. C	17. D	18. A
19. D	20. B	21. B
22. A	23. C	24. D
25. A	26. A	27. C
28. C	29. B	30. D
31. C	32. D	33. C
34. B	35. A	36. D
37. A	38. B	39. B
40. B	41. C	42. D
43. C	44. B	45. A
46. C	47. C	48. D
49. B	50. A	51. A
52. B	53. C	54. C
55. B	56. D	57. D
58. B	59. C	60. D
61. A	62. B	63. D
64. D	65. B	66. B
67. B	68. A	69. A
70. B	71. C	72. A
73. C	74. D	75. A

Journeyman-Level CET Practice Test
Communications option

Please read this first! One important difference is that special hints for preparation are given in this practice test. You will find a discussion of what is to be expected for each section.

In a Communications CET test, you might get questions on AM, FM, sideband, transmitters, receivers, and transceivers. Also, you can expect questions on basics with emphasis on tests, measurements, and troubleshooting. In comparison with the FCC tests, you can expect less emphasis on broadcasting and more emphasis on two-way radio. Also, there are no questions on rules and regulations in the CET test.

The practice test starts with Section IX. The titles of sections that differ from the actual exam that you receive either from ETA or ISCET. Refer to Appendix F at the end of this book for more on ETA and ISCET Journeyman Communications CET Exams.

Be prepared to answer questions on tubes, bipolar transistors, and field-effect transistors. It is especially important to know the polarities of voltages needed to get these devices into operation. Example: In an npn transistor should the base be positive or negative with respect to the collector? Answer: Negative.

Circuits using these devices are included. Know the advantages and disadvantages of each basic configuration. Example: Which FET configuration has a high input impedance and low output impedance? Answer: Common drain, which is also known as a source follower.

Be able to identify amplifier configurations. For example, know the types of power amplifiers, such as totem poles, quasi-complementary, and parallel-operated. In questions on amplifiers, it is important to be able to follow signal paths. Know when a signal is inverted and when it is not inverted.

Be able to identify the different oscillator circuits, and whether an oscillator is series or shunt fed. Crystal oscillators are especially important in this type of test.

Questions on filters, tuned circuits, and series-parallel combinations of components are usually given in the Associate-Level test. However, it is not unusual to have a few questions on basics here. You will find examples in this practice test.

76. Regarding the device shown in Fig. 9-1, and with the dc voltages shown, which of the following statements is correct?

   A. The device is conducting the maximum possible drain current, and may be destroyed.
   B. The device is a P-channel enhancement MOSFET.
   C. The input signal is normally delivered to the drain.
   D. None of these choices is correct.

77. Which of the following equations can be used for finding the wavelength of a radio wave?

   A. Wavelength in meters.
   B. Wavelength in miles.
   C. Both equations are correct.
   D. Neither equation is correct.

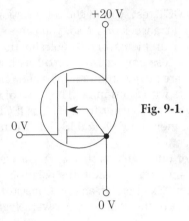

Fig. 9-1.

78. The circuit in Fig. 9-2 can be used as:

    A. a wavemeter.
    B. a Maxwell bridge.
    C. an RF wattmeter.
    D. a local oscillator.

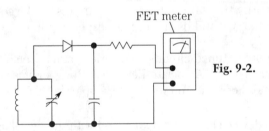

Fig. 9-2.

79. Figure 9-3 shows two rectifier diodes in series. The purpose of the resistors is to:

    A. limit current surges caused by the charging filter capacitor.
    B. equalize voltage drops across the diodes.
    C. provide additional filtering for low-frequency ripple.
    D. None of these answers is correct.

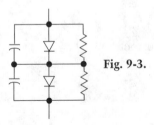

Fig. 9-3.

80. The purpose of the bypass capacitors in Fig. 9-3 is to:

    A. prevent inductive kickback damage from the ripple filter.
    B. increase the junction capacitance of the diodes.
    C. prevent damage to the diodes from transient spikes.
    D. filter low-frequency ripple.

81. Which of the following components could be used in place of C in the circuit of Fig. 9-4?

    A. Forward-biased hot-carrier diode.
    B. Reverse-biased silicon junction diode.
    C. Forward-biased PIN diode.
    D. Reverse-biased germanium point-contact diode.

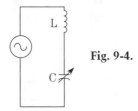

Fig. 9-4.

82. In the circuit of Fig. 9-4, the capacitance of C is varied by moving the plates closer together or further apart. In order to increase the resonant frequency you should:

    A. move the capacitor plates further apart.
    B. move the capacitor plates closer together.

83. Undesired, high-frequency oscillations in an amplifier are called:

    A. parasitics.
    B. transients.
    C. VCOs.
    D. flutters.

84. A certain resistor is color coded brown, black, black, and gold. The highest resistance this resistor can have and still be in tolerance is:

    A. 105 Ω.
    B. 1.05 Ω.
    C. 10.5 Ω.
    D. None of these answers is correct.

85. For the op-amp circuit of Fig. 9-5:

    A. the output signal is 180° out of phase with the input signal.
    B. the gain is $R_3/R_1 + R_2$
    C. the terminal marked with a + sign is never used for input signal.
    D. the terminal marked with a – sign shows when the negative side of the power supply is connected.

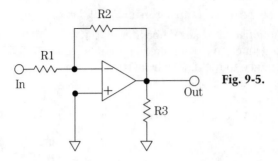

Fig. 9-5.

86. The power factor of the circuit in Fig. 9-6 will have a value of 1.0 when the switch is in position:

    A. x.
    B. y.
    C. z.
    D. None of these answers is correct.

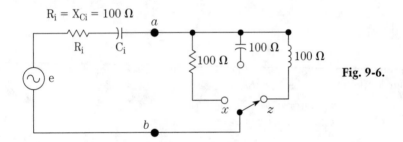

Fig. 9-6.

87. In the circuit of Fig. 9-6, the ac generator terminals are a and b. Ri and Ci are the internal resistance and reactance of the generator. The generator will deliver the most power to the load when the switch is in position:

    A. x.
    B. y.
    C. z.
    D. None of these answers is correct.

88. The transistor in the oscillator circuit of Fig. 9-7 is operated:

    A. Class A.
    B. Class B.
    C. Class C.
    D. Cannot determine from information given.

89. The power supply voltage in the circuit of Fig. 9-7 should be:

    A. positive.
    B. negative.

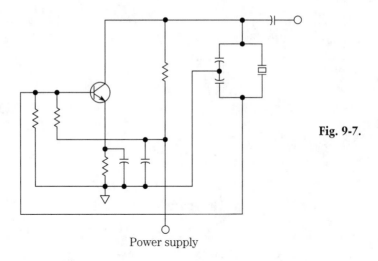

Fig. 9-7.

Power supply

90. The transistor oscillator in the circuit of Fig. 9-7 is:

    A. series fed.
    B. shunt fed.

91. The transistor in the circuit of Fig. 9-7 is connected in the:

    A. common-base configuration.
    B. common-emitter configuration.

92. In Fig. 9-8, the resonant frequency of the parallel L-C circuit is 100 kHz. As seen by the 110-kHz ac generator, the circuit is:

    A. open.
    B. resistive.
    C. capacitive.
    D. inductive.

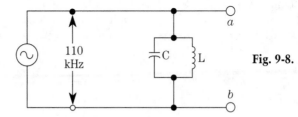

Fig. 9-8.

93. Adding a resistor between terminals a and b in the circuit of Fig. 9-8 will:

    A. make the circuit tune more like y in Fig. 9-9.
    B. make the circuit tune more like x in Fig. 9-9.

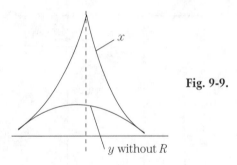

Fig. 9-9.

94. Greater "talk power" is obtained in a transmitter by using:

    A. a speech clipper.
    B. an equalizer.
    C. a carbon microphone with dc bias.
    D. slightly greater than 100% modulation.

95. Regarding receiver operation, the initials BFO stand for:

    A. Bare Foot Operation.
    B. Beat Frequency Oscillator.
    C. Break Front Operation.
    D. Backward-Forward Oscillation.

In addition to transporting a signal from one point to another, transmission line segments are used for impedance matching, tuning, mechanical support, and frequency measurement.

It would be a good idea to know the characteristics of both open and shorted lines, at various lengths up to and including half wave. Standing waves, and the measurement of standing wave ratios should be clearly understood. Impedances of transmission lines are also the subject of some questions in this section.

The methods of matching transmission lines to antennas are important. Also, the types of antennas, especially the quarter-wave and half-wave types, must be known. The use of loading capacitors and loading coils to change the electrical length of antennas will be useful information for this part of the test.

96. Another name for the characteristic impedance of a transmission line is:

    A. bode impedance.
    B. surge impedance.
    C. I^2R impedance.
    D. $E\text{-}I$ impedance.

97. Which of the following statements is true?

    A. Above 10 MHz, ground-wave coverage is greatly reduced.
    B. Above 10 MHz, ground-wave coverage increases with frequency.

98. Which of the following standing-wave ratios is more desirable?

    A. 1:1
    B. 10:1.

99. The base of a vertical quarter-wave antenna that is shunt fed:

    A. must never be grounded.
    B. can be grounded.

100. A quarter-wave stub, with a shorted end, has:

    A. nearly zero impedance at the open end.
    B. an extremely high impedance at the open end.

101. The tendency for high-frequency currents to flow near the surface of conductors, rather than through the entire cross section is called:

    A. skin effect.
    B. outer flow ratio.
    C. inverse resistance ratio.
    D. peel effect.

102. You could use a grid-dip meter (or semiconductor equivalent) to measure:

    A. undesired radiated power from a transmission line.
    B. IC voltages on dual-inline packages.
    C. grid modulation percentage.
    D. the resonant frequency of an L-C tank circuit.

103. Lecher lines are used for measuring:

    A. frequency.
    B. radiation resistance.
    C. power.
    D. efficiency.

104. If you cut a 100' section of 72-Ω coaxial cable into two 2' lengths, each section will have an impedance of:

    A. 36 Ω.
    B. 72 Ω.
    C. 144 Ω.
    D. 300 Ω.

105. In the transformer symbol of Fig. 10-1, the purpose of the shield (marked X) is to:

    A. reduce the primary current.
    B. reduce hysteresis losses.
    C. reduce eddy current losses.
    D. prevent electrostatic coupling between the primary and secondary.

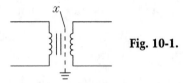

Fig. 10-1.

106. Which of the antennas in Fig. 10-2 will radiate a wave with a horizontal magnetic field?

    A. The one marked B.

    B. The one marked A.

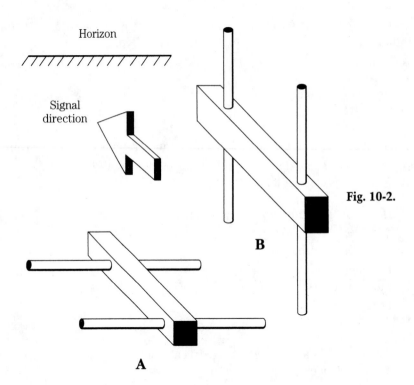

Fig. 10-2.

107. If a piece of wire is too short to be used as a quarter-wave antenna, it can be made electrically longer by adding:

    A. inductive loading.

    B. capacitive loading.

108. Devices for matching a balanced transmission line and load to an unbalanced transmission line and generator, with maximum transfer of power between the two, are called:

  A. reeds.
  B. diacs.
  C. baluns.
  D. Z-angle matches.

109. Which of the following is not an example of transmission line loss?

  A. Hysteresis loss.
  B. Dielectric loss.
  C. Copper loss.
  D. Radiation loss.

110. Which of the following antennas most nearly approaches an isotropic radiator?

  A. Loop.
  B. Rhombic.
  C. Marconi.
  D. Hertz.

111. A director in an antenna system is an example of:

  A. a shield.
  B. a parasitic element.
  C. a driven element.
  D. an array.

112. With a diversity antenna system, the receiving antennas should be:

  A. not less than one wavelength apart.
  B. more than one-half wavelength, but less than a full wavelength apart.
  C. one-half wavelength apart.
  D. one-fourth wavelength apart.

113. Which of the following is not a method of matching a transmission line to an antenna?

  A. Transmission line stub.
  B. Gamma match.
  C. Delta match.
  D. Kappa match.

114. As shown in Fig. 10-3, a 20' length of 72-Ω coaxial cable is terminated with a pure resistance (no inductance or capacitance). Assuming that the generator is matched to the line, the resistor will dissipate the maximum possible heat when the resistance value is:

  A. $20 \times 72 = 1440 \ \Omega$.
  B. 72 Ω.
  C. 36 Ω.
  D. 0 Ω.

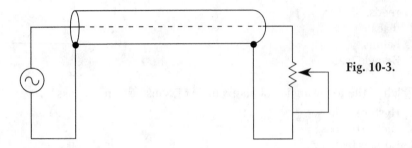

Fig. 10-3.

115. Regarding a Marconi antenna, which of the following is true?

    A. It is no longer being used.
    B. Voltage is maximum at the base, minimum at the top.
    C. Current is maximum at the base, minimum at the top.
    D. It is a half-wave, center-fed antenna.

Questions in this section are mostly related to AM, FM, and single-sideband operation. There might be an occasional question on pulse-position modulation. However, unlike the FCC test, there are no questions on television modulation or demodulation. Know the Hazeltine, Rice, and lower-level vs. high-level modulation circuits.

It is very important to know the measurements related to modulation. This includes such factors as percent modulation, modulation index, deviation, and sideband frequencies.

All types of demodulators are included in questions. Know the difference between ratio detectors and discriminators, and other FM detectors.

Phase-locked loops (PLLs) are becoming increasingly important because of their use as detectors, and also because of their use in frequency synthesizers. Part of this popularity is caused by the availability of integrated circuit PLLs, such as the 565. It would be a good idea to know the basics of PLLs for this section and for other sections of the CET test.

Superheterodyne receivers are most popular, but other types should also be understood. Remember that the converter in a superhet is sometimes called the *first detector*, and this means that you might get questions on image frequencies, frequency conversion, and double conversion.

116. When a pure sine wave with a frequency of 1200 Hz is used to modulate a 27.5-MHz signal in an AM transmitter, the lower sideband frequency is:

    A. 26.3 MHz.
    B. 26.12 MHz.
    C. 27.5 MHz.
    D. None of these answers is correct.

117. A downward carrier shift with modulation in an AM transmitter:
   A. means that the negative peaks are greater than the positive peaks.
   B. means that the frequency of the carrier goes down when modulation is applied.

118. A certain superheterodyne receiver has a 455-kHz IF frequency. When tuned to a 1500-kHz signal, one of the image frequencies will be:
   A. 3000 kHz.
   B. 2410 kHz.
   C. 1545.5 kHz.
   D. 900 kHz.

119. The circuit in Fig. 11-1 is a:
   A. high-pass filter.
   B. band-rejection filter.
   C. low-pass filter.
   D. band-pass filter.

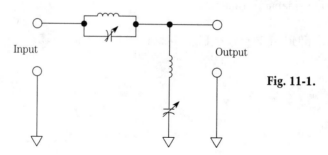

Input   Output

Fig. 11-1.

120. Which of the following might be used as a local oscillator in the TRF receiver?
   A. Armstrong.
   B. Shunt-fed Hartley.
   C. Both of these may be used.
   D. Neither of these could be used.

121. To eliminate unnecessary noise from a receiver when it is not receiving a signal, the receiver might use a:
   A. squelch circuit.
   B. reducer.
   C. noise gate circuit.
   D. noise limiter circuit.

122. In a broadcast FM signal, the bandwidth increases with:

    A. an increase in the amplitude of the audio modulating signal.

    B. an increase in the frequency of the audio modulating frequency.

123. A voltage-controlled oscillator:

    A. could not be used in a direct FM transmitter to generate the carrier.

    B. could be used in a direct FM transmitter to generate the carrier.

124. In comparing typical Hartley oscillators:

    A. both tube and bipolar transistor circuits are operated class B.

    B. both tube and bipolar transistor circuits are operated class A.

    C. both tube and bipolar transistor circuits are operated class C.

    D. those using tubes are operated class C, while bipolar transistor types are forward biased for class-AB operation.

125. You would expect to find a de-emphasis circuit in:

    A. an FM transmitter.

    B. an FM receiver.

    C. a single-sideband transmitter.

126. In the modulation display of Fig. 11-2, $A = 2.5$ cm and $B = 3.5$ cm. The percent modulation is at:

    A. 41⅔%.

    B. 28.57%.

    C. 71.4%.

    D. 16⅔%.

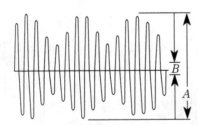

Fig. 11-2.

127. A balanced modulator is used in single-sideband transmitters because:

    A. it automatically eliminates the carrier.

    B. it has no phase distortion.

    C. it can be made with one pnp and one npn transistor.

    D. it automatically eliminates one sideband.

128. The oscillator shown in Fig. 11-3 is:

    A. an Armstrong type.

    B. a Pierce type.

    C. a Colpitts type.

    D. an R-C phase shift type.

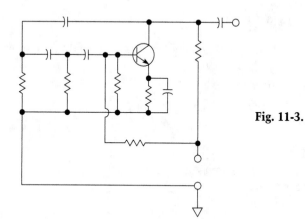

Fig. 11-3.

129. If the input signal of the frequency multiplier in Fig. 11-4 is off frequency by 0.05%, the error in output frequency will be:

    A. 23.25 kHz.

    B. 19.375 kHz.

    C. 0.0316 MHz.

    D. 0.018 MHz.

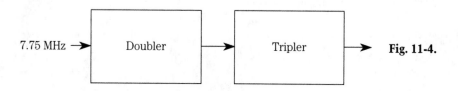

Fig. 11-4.

130. Consider this equation for an FM signal:

$$Deviation\ ratio = Deviation \backslash Modulating\ frequency$$

    A. The equation is not correct.

    B. The equation is correct.

131. The output frequency for the frequency multiplier system in Fig. 11-4 should be:

    A. 12.75 MHz.

    B. 46.5 MHz.

    C. 38.75 MHz.

    D. None of these answers is correct.

132. Consider this equation for an MF signal:.

Frequency modulation deviation × Modulation frequency = index

    A. This equation is not correct.

    B. This equation is correct.

133. Which of the following FM demodulators is preceded by a limiter stage?

    A. Gated beam detector.

    B. Discriminator.

    C. Phase-locked loop.

    D. Ratio detector.

134. Splattering is caused by:

    A. broken parasitic elements on an antenna.

    B. oscillator pulling.

    C. excessive modulation.

    D. insufficient modulation.

135. Which of the following is true regarding the trapezoid pattern of Fig. 11-5?

    A. It is used only for determining the FM modulation index.

    B. It can be used for calculating percent modulation of an AM signal using the following equation:.

$$\% \ Modulation = x/y \times 100$$

    C. It shows that the output signal is greatly distorted.

    D. None of these statements is correct.

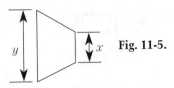

Fig. 11-5.

    Technicians usually do well with this section. It includes circuits and components used in transmitters, receivers, and transceivers. As in other sections, troubleshooting and measurements are emphasized.

    An important part of this section is the test of your ability to work directly in circuits. There are examples in this practice test (Questions 136 to 142). It is best to study the complete circuit board first to get an idea of the circuitry. This approach is better than starting to answer questions about parts of the circuit without having an overall picture. Get the input and output terminals clear in your mind, and various points for making measurements.

Be sure to read each question carefully, and be sure you answer the question being asked. Interviews with technicians after they have taken the CET test reveal that carelessness is an important factor in the number of questions answered incorrectly.

Figure 12-1 shows the component layout for a regulated power supply. Questions 136 to 142 test your ability to analyze a circuit from components, rather than from a schematic diagram. In this practice test, the schematic for the circuit is shown in Fig. 12-1. The schematic would not normally be given in the actual CET test!

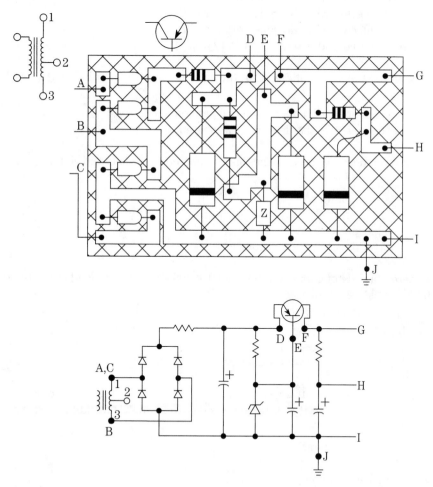

Fig. 12-1.

136. To make a bridge rectifier in the circuit of Fig. 12-1:
    A. connect leads A and B to transformer terminal #2, and lead C to transformer terminal #1.
    B. connect lead A to transformer terminal #1, lead B to transformer terminal #2, and lead C to transformer terminal #3.
    C. connect leads A and B to transformer terminal #1, and lead C to transformer terminal #3.
    D. connect leads A and C to transformer terminal #1, and lead B to transformer terminal #3.

137. Regarding the circuit of Fig. 12-1, the pnp power transistor must:
    A. be connected with the collector at lead F.
    B. be connected with the collector at lead E.
    C. be connected with the collector at lead D.
    D. not be used in this positive power supply.

138. Regarding the circuit of Fig. 12-1, the cathode of the zener diode (Z) should be connected to:
    A. the copper strip that leads I and J are connected to.
    B. the copper strip that lead E is connected to.

139. There is a resistor between the foils connected to leads G and H in Fig. 12-1. It is used:
    A. as a surge limiter.
    B. for obtaining bias.
    C. as a bleeder.
    D. None of these answers is correct.

140. There is a resistor connected between the foil connected to lead D and the rectifier. This resistor is used as a:
    A. surge-limiting resistor.
    B. bias resistor.
    C. voltage dropping resistor.
    D. None of these answers is correct.

141. Assume that a power transistor is mounted outboard in the circuit of Fig. 12-1. To increase the current rating of the regulated power supply, an additional transistor can be added in:
    A. an AND configuration.
    B. a Darlington configuration.
    C. a stacked amplifier configuration.
    D. a totem pole configuration.

142. A ferrite bead is to be added to the circuit of Fig. 12-1. Its purpose is to prevent parasitic oscillations. It would normally be added to:

A. Lead I.
B. Lead E.
C. Lead B.
D. Lead A.

143. Which of the following stages might be used to follow the oscillator that generates the carrier in an AM transmitter?

A. Low-pass filter.
B. Inverter/converter.
C. AGC/AFC.
D. Buffer/multiplier.

144. A number of different frequencies for a transmitter carrier can be obtained
with two crystals and:

A. a synthesizer.
B. a Loftin-White amplifier.
C. a digital/analog converter.
D. an op-amp.

145. Which of the following is a circuit used for neutralizing an amplifier?

A. Rice.
B. Parametric.
C. Phase lock.
D. Snubber.

146. Which of the following amplifiers has a high input impedance and low output
impedance?

A. Bootstrap amplifier.
B. Common-collector amplifier.
C. Common-base amplifier.
D. Common-emitter amplifier.

147. Figure12-2 shows the response curve of a tuned circuit. The bandwidth for this
curve is:

A. measured at the half power points.
B. equal to the center frequency divided by the circuit Q.
C. measured at the points where the amplitude is down to 0.707 of the maxi-
mum.
D. All of these answers are correct.

148. In the circuit of Fig. 12-3, the phase angle between the voltage and current
would:

A. increase if the plates of capacitor C were moved further apart.
B. increase if the plates of capacitor C were moved closer together.

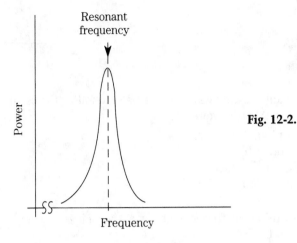

Fig. 12-2.

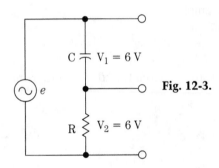

Fig. 12-3.

149. Which of the following statements is true regarding the circuit in Fig. 12-3?

   A. The value of e cannot be obtained by adding V_1 and V_2.
   B. $e = V_1 + V_2 = 12$ V.

150. A dc voltage is not normally required for the operation of a:

   A. carbon microphone.
   B. ribbon microphone.
   C. "condenser" microphone.
   D. All of these answers are correct.

Answer sheet for Journeyman-Level Practice CET Test

76. D	77. C	78. A
79. B	80. C	81. B
82. A	83. A	84. C
85. A	86. C	87. A
88. A	89. A	90. B

91. B	92. C	93. A
94. A	95. B	96. B
97. A	98. A	99. B
100. B	101. A	102. D
103. A	104. B	105. D
106. A	107. A	108. C
109. A	110. C	111. B
112. A	113. D	114. B
115. C	116. D	117. A
118. B	119. B	120. D
121. A	122. A	123. B
124. D	125. B	126. D
127. A	128. D	129. A
130. A	131. B	132. A
133. B	134. C	135. D
136. D	137. A	138. B
139. D	140. A	141. B
142. B	143. D	144. A
145. A	146. B	147. A
148. B	149. A	150. B

A
APPENDIX

Diode color codes

The JEDEC number has a "1N" with a sequence number of four digits. The four digits are given by four color bands. The first digit is a broad band that also indicates the cathode end. The same color code as used for resistors applies.

The Pro Electron number has three letters and a sequence of two digits. The letters are indicated by two broad bands at the cathode end. The digits are shown by small bands. The color code is given in Table A-1. Also see Fig. A-1.

Table A-1.

Broad bands		Small bands
First band	**Second band**	**Serial number**
AA-brown	Z-white	0-black
BA-red	Y-grey	1-brown
	X-black	2-red
	W-blue	3-orange
	V-green	4-yellow
	T-yellow	5-green
	S-orange	6-blue
		7-violet
		8-grey
		9-white

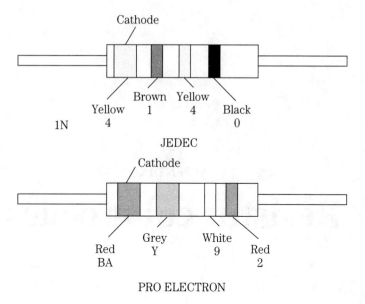

1N

JEDEC

PRO ELECTRON

Fig. A-1.

Resistor color codes

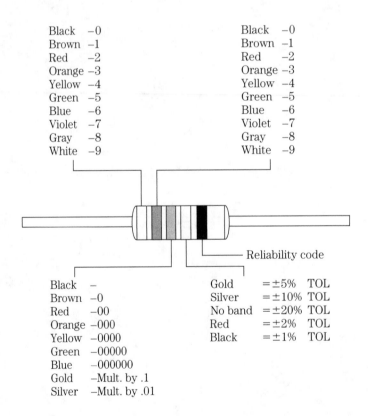

Black   –0
Brown   –1
Red   –2
Orange –3
Yellow –4
Green   –5
Blue   –6
Violet –7
Gray   –8
White   –9

Black   –0
Brown   –1
Red   –2
Orange –3
Yellow –4
Green   –5
Blue   –6
Violet –7
Gray   –8
White   –9

Reliability code

Black   –
Brown   –0
Red   –00
Orange –000
Yellow –0000
Green   –00000
Blue   –000000
Gold   –Mult. by .1
Silver –Mult. by .01

Gold   =±5%   TOL
Silver   =±10% TOL
No band =±20% TOL
Red   =±2%   TOL
Black   =±1%   TOL

Radial lead dot resistor

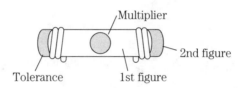

Radial lead (band) resistor

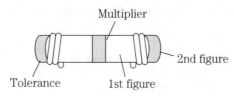

Examples:
Color Code: Red, Violet, Orange, Gold
27,000 Ω ±5%
Color Code: Brown, Black, Black, Gold
10 Ω ±5%

Body-dot system

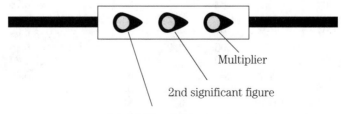

Dash-band system

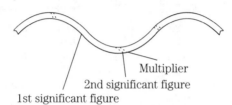

Miniature resistor code

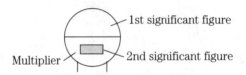

Dot-band system

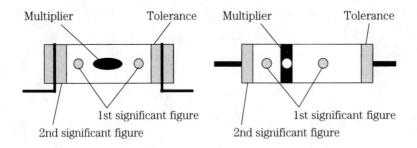

Standard resistance decade value

E 48 (±2%)	E 96 (±1%)	E 48 (±2%)	E 96 (±1%)	E 48 (±2%)	E 96 (±1%)	E 48 (±2%)	E 96 (±1%)	E 24 (±5%)
1.00	1.00	1.78	1.78	3.16	3.16	5.62	5.62	1.0
	1.02		1.82		3.24		5.76	1.1
1.05	1.05	1.87	1.87	3.32	3.32	5.96	5.90	1.2
	1.07		1.91		3.40		6.04	1.3
1.10	1.10	1.96	1.96	3.48	3.48	6.19	6.19	1.5
	1.13		2.00		3.57		6.34	1.6
1.15	1.15	2.05	2.05	3.65	3.65	6.49	6.49	1.8
	1.18		2.10		3.74		6.65	2.0
1.21	1.21	2.15	2.15	3.83	3.83	6.81	6.81	2.2
	1.24		2.21		3.92		6.98	2.4
1.27	1.27	2.26	2.26	4.02	4.02	7.15	7.15	2.7
	1.30		2.32		4.12		7.32	3.0
1.33	1.33	1.33	2.37	2.37	4.22	7.50	7.50	3.3
	1.37		2.43		4.32		7.68	3.6
1.40	1.40	2.49	2.49	4.42	4.42	7.87	7.87	3.9
	1.43		2.55		4.53		8.06	4.3
1.47	1.47	2.61	2.61	4.64	4.64	8.25	8.25	4.7
	1.50		2.67		4.75		8.45	5.1
1.54	1.54	2.74	2.74	4.87	4.87	8.66	8.66	5.6
	1.58		2.80		4.99		8.87	6.2
1.62	1.62	2.87	2.87	5.11	5.11	9.09	9.09	6.8
	1.65		2.94		5.23		9.31	7.5
1.69	1.69	3.01	3.01	5.36	5.36	9.53	9.53	8.2
	1.74		3.00		5.49		9.76	9.1

C
APPENDIX

Capacitor color codes

Molded paper capacitor codes
(capacitance given in pF)

Color	Digit	Multiplier	Tolerance
Black	0	1	20%
Brown	1	10	
Red	2	100	
Orange	3	1000	
Yellow	4	10000	
Green	5	100000	5%
Blue	6	1000000	
Violet	7		
Gray	8		
White	9		10%
Gold			5%
Silver			10%
No color			20%

Molded capacitor color codes

5-dot radial-lead
ceramic capacitor

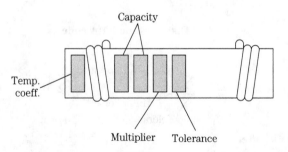

Extended-range
TC ceramic hicap

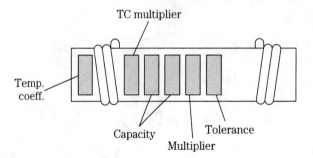

Axial lead
ceramic capacitor

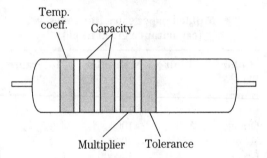

By-pass coupling
ceramic capacitor

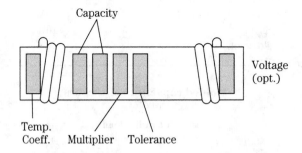

Disc ceramic RMA code

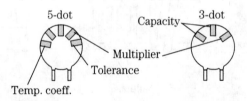

Mica capacitor color code

Color	Characteristic* (EIA)	Capacitance 1st and 2nd significant figures	Multiplier	Capacitance tolerance	dc working voltage	Operating temperature range	Vibration grade (Mil)
Black	A (EIA)	0	1	±20% (EIA)		−55° to +70°C (Mil)	10–55 Hz
Brown	B	1	10	±1%	100 (EIA)		
Red	C	2	100	±2%		−55° to +85°C	
Orange	D	3	1000		300		
Yellow	E	4	10000 (EIA)			−55° to +125°C	10–2000 Hz
Green	F	5		±5%	500		
Blue		6				−55° to +150°C (Mil)	
Purple (violet)		7					
Gray		8					
White		9					
Gold			0.1	±½% (EIA)†	1000 (EIA)		
Silver			0.01 (EIA)	±10%			

* Denotes specifications of design involving Q factors, temperature coefficients, and production test requirements

† Or ±0.5 pF, whichever is greater. All others are specified tolerance or ±1.0 pF, whichever is greater.

Mica capacitor color codes
Current standard code

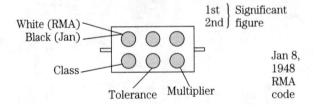

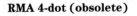

RMA 3-dot (obsolete)
Rated 500 W.V.D.C. ±20% Tol.

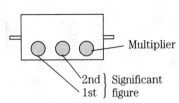

RMA 4-dot (obsolete)

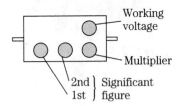

Button silver mica capacitor

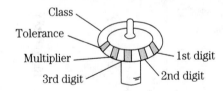

RMA (5-dot obsolete code)

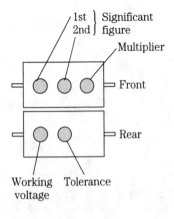

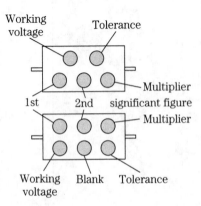

RMA 6-dot (obsolete)

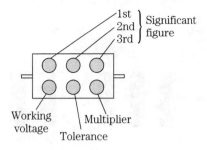

Tubular capacitor color codes

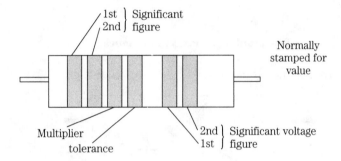

A 2-digit voltage rating indicates
more than 900 V.
Add 2 zeros to end of 2-digit number.

Molded flat capacitor commercial code

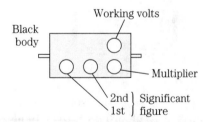

Jan. code capacitor

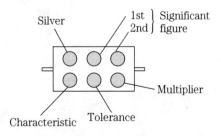

Molded ceramics
Using standard resistor color code

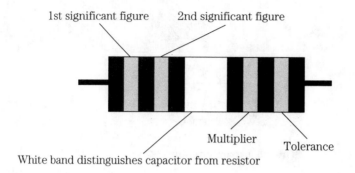

1st significant figure

2nd significant figure

Multiplier

Tolerance

White band distinguishes capacitor from resistor

Typographically marked ceramics

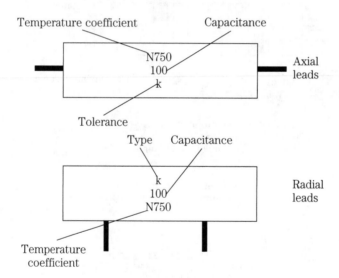

Temperature coefficient

Capacitance

N750
100
k

Axial leads

Tolerance

Type   Capacitance

k
100
N750

Radial leads

Temperature coefficient

Tolerance

Letter	10 pF or less	Over 10 pF
B	±0.1 pF	
C	±0.25 pF	
D	±0.5 pF	
F	±1.0 pF	±1%
G	±2.0 pF	±2%
J		±5%
K		±10%
M		±20%

Ceramic capacitor codes
(capacitance given in pF)

In reference to RS-198-A of the EIA, ceramic dielectric capacitors have three major classifications:

Class 1, temperature compensating ceramics requiring high Q and capacitance stability.

Class 2, where Q and stability of capacitance are not required.

Class 3, low voltage ceramics, where dielectric losses, high insulation resistance, and capacitance stability are not of major importance.

| | | | *Tolerance** | | Class 1 | | Coeff. | Class 2 |
Color	Digit	Multiplier	10 pF or less	Over 10 pF	Temperature coefficient PPM°C	Temp. Significant figure	Multiplier	Tolerance*
Black	0	1	±2.0 pF	±20%	0	0.0	−1	±20%
Brown	1	10	±0.1 pF	±1%	−33		−10	
Red	2	100		±2%	−75	1.0	−100	
Orange	3	1000		±3%	−150	1.5	−1000	
Yellow	4	10,000			−220	2.2	−10,000	+100%, −0%
Green	5		±0.5 pF	±5%	−330	3.3	+1	±5%
Blue	6				−470	4.7	+10	
Violet	7				−750	7.5	+100	
Gray	8	.01	±0.25 pF		+150 to −1500		+1000	+80%, −20%
White	9	.1	±1.0 pF	±10%	+100 to −750		+10,000	±10%
Silver								
Gold								

*Tolerance on class-3 ceramic capacitors is indicated by its code, either ±20% (code M) or +80, −20% (code Z).

Extended-range T.C. tubular ceramics

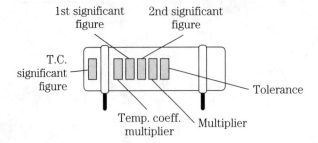

Feedthrough ceramics

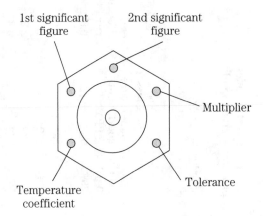

Button ceramics

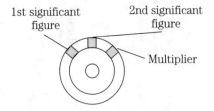

Viewed from soldered surface

Standoff ceramics

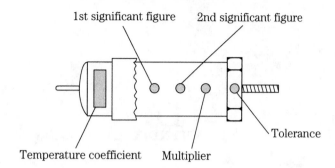

1st significant figure   2nd significant figure

Tolerance

Temperature coefficient   Multiplier

Temperature compensating
tubular ceramics

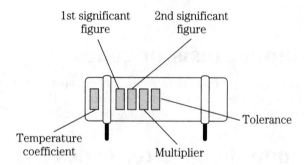

1st significant
figure

2nd significant
figure

Tolerance

Temperature
coefficient   Multiplier

Surface-mount color codes

Surface-mount resistor codes

No color codes are used for surface-mount resistors or capacitors. However, the values are marked in alphanumeric examples. For resistors, 473 and 47K would both represent 47 kΩ. A 4.7 kΩ resistor would be marked 4K7.

Surface-mount capacitor codes

For capacitors, letters are used to represent the values. For example, E3 represents 1.5×10^3 picofarads. That value can also be written as 1K3.

A = 1.0	J = 2.2	S = 4.7	a = 2.5
B = 1.1	K = 2.4	T = 5.1	b = 3.5
C = 1.2	L = 2.7	U = 5.6	d = 4.0
D = 1.3	M = 3.0	V = 6.2	f = 5.0
E = 1.5	N = 3.3	W = 6.8	m = 6.0
F = 1.6	P = 3.6	X = 7.5	n = 7.0
G = 1.8	Q = 3.9	Y = 8.2	t = 8.0
H = 2.0	R = 4.3	Z = 9.1	y = 9.0

E
APPENDIX

Abbreviations used in FCC and CET tests

AAB	Automatic Answer Back
ADC or *A/D*	Analog-to-Digital Converter
AF	Audio Frequency
AFC	Automatic Frequency Control
AM	Amplitude Modulation
ARPA	Automatic Radar Plotting Aid
ARQ	Automatic Repeat Request
ASCII	American Standard Code Information Interchange
AGC	Automatic Gain Control
BFO	Beat Frequency Oscillator
CCITT	(In English) Consulting Committee for International Telephone and Telegraph
CMOS	Complementary Metal Oxide Semiconductor
CODEC	COder DECoder
COLE (m)	Commercial Operator Licensing Examination (Manager)
CQ	Seek You
CTCSS	Continuous Communications Tone-Coded Squelch System
dB	deciBels
DME	Distance Measuring Equipment
DSC	Double Sideband with Carrier
DTMF	Dual-Tone Multi Frequency
ECL	Emitter Coupled Logic
EIRP	Effective Isotropic Radiated Power

EPIRB	Emergency Position indicating Radio Beacon
FCC	Federal Communications Commission
FET	Field-Effect Transistor
FM	Frequency Modulation
FSK	Frequency Shift Keying
GMDSS	Global Marine Distress and Safety System
HF	High Frequency
IF	Intermediate Frequency
IMO	International Maritime Organization
IOS	International Organization for Standardization
ITU	International Telecommunications Commission
kHz	kiloHertz
LED	Light Emitting Diode
LORAN	LOng RANge (Navigation System)
LSB	Least Significant Bit
MF	Medium Frequency
MSB	Most Significant Bit
MOSFET	Metal-Oxide Semiconductor
MROP	Marine Radio Operator Permit
Op Amp	Operational Amplifier
PAM	Pulse Amplitude Modulation
PCM	Pulse Code Modulation
PD (or DP)	Potential Difference (or, Difference of Potential)
PEP	Peak Envelope Power
PLL	Phase-Locked Loop
PPM	Pulse-Position Modulation
PPC	Proof-of-Passing Certificate
PWM	Pulse-Width Modulation
Q Signals	See Appendix III
RADAR	Radio Detection and Ranging
RF	Radio Frequency
RISC	Reduced Instruction Set Computer
RTTY	Radio Teletype
SAW (filter)	Surface Audio Wave (filter)
SCA	Subsidiary Communications Authorization
SCR	Silicon Controlled Rectifier
SF	Signal Frequency
SINAD	Signal + Noise and Distortion
SMSA	Ship Movement Safety Agency
SSB	Single Side Band
SSSC	Signal Sideband Suppressed Carrier
STC	Sensitivity Time Control
SWR	Standing Wave Ratio
TDM	Time Division Multiplier
TDR	Time Delay Reflectometry
TE	Transverse Electric (waveguide mode)

TM	Transverse Magnetic (waveguide mode)
TRF	Tuned Radio Frequency (receiver)
TTL	Transistor Transistor Logic
UTC	Universal Time-Coordinated
VCO	Voltage Controlled Oscillator
VHF	Very High Frequency
VIRS	Vertical Interval Reference Signal
VITS	Vertical Interval Test Signed
VMOS	Vertical Metal Oxide Semiconductor (The V refers to the method of Fabrication)
VSWR	Voltage Standing Wave Ratio
VTS	Vessel Traffic Service
WSMB	Water Safety Management Bureau
WVDC	Working Voltage DC
XNOR	Exclusive NOR
XOR	Exclusive OR

Present ETA and ISCET Journeyman CET exams

ETA Exams

The present ETA Journeyman Communications CET Exams are available as either a Radio Option or a Telecommunications Option. The Radio Journeyman Communications Option covers 2-way, amateur, and business radio; maritime, aviation, and naviation radio; communications, walkie talkies, and short-range radio; military radio and radar. The Radio Journeyman Telecommunications Option includes television, telephone, and TV distribution, cellular phones and pagers, data, microwave, and fiberoptic communications.

ISCET Exams

The present ISCET Journeyman Communications CET Exam includes four sections: Basic Communications Electronics, 25 questions; AM and FM Transmitters, 20 questions; Communications Receivers, 20 questions; and Communications Systems, 20 Questions. Covered are antennas and transmission lines; Propagation, filters, oscillators; frequency and time division; two-way radio, frequency tolerance, deviation, and modulation; microwaves, communications receivers, and single sideband; two-way communications, telemetry, and cellular radio. ISCET also has Journeyman CET tests on the Radar Electronics Option, the Video Option, the Audio Option, and FCC Legal.

Contact the Associations for further information at the following addresses:

Electronic Technicians Association (ETA)
602 North Jackson
Greencastle, Indiana 46135
317-653-8262

International Society of Certified Electronic Technicians (ISCET)
2708 West Berry Street
Fort Worth, Texas 76109
817-921-9061 or 817-921-9101

Index

Illustrations are in **boldface**.

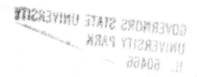

About the authors

Sam Wilson

Sam Wilson earned his Bachelor's degree from Long Beach State College and his Master's degree from Kent State University. He also has diplomas from Capitol Radio Engineering Institute and RCA Institutes.

Wilson is now a full-time technical writer and consultant. In 1983, he was selected as *Technician of the Year* by the International Society of Certified Electronics Technicians (ISCET). He has been the CET Test Consultant for that organization; and, has been the Technical publications Director for the National Electronics Service Dealers Association (NESDA).

His electronics experience includes 18 years as an instructor and professor. He also has 12 years of practical experience as a technician and engineer. He is presently a specialist in training equipment design and preparation of technical training publications. Over the years, Wilson has written over 25 technical electronics books.

Joe Risse

Joseph A. "Joe" Risse has worked as an assembler, technician, engineer, maintenance engineer, transmitter operator, chief broadcast engineer, director of electronics department for a correspondence school, project manager for industrial training programs, and is active on advisory committees for a technical institute and a vo tech school. He has completed courses in the military, college courses in electronics, correspondence school courses, and industrial group/training programs.

He holds the B.A. degree in Natural Science/Mathematics from Thomas Edison College, is a Fellow of the Society of Broadcast engineers and the Radio Club of America, a Life Member of the Electronic Technicians Association, and Member of the International Society of Certified Electronics Technicians. He is a registered Professional Engineer by the Commonwealth of Pennsylvania, and certified as an electronics technician by both ISCET and ETA.

Risse has completed all of the electronics, mathematics, and physics courses required for the Electronics major in Physics degree at the University of Scranton. He has prepared certification and practice exams for ISCET. He was the editor of the Journal of the Society of Broadcast Engineers during the early years of the SBE, and held the national office of executive vice president for 2 years.